ALLE ZEIT WACH 1842

O. Sild K. Haller (Eds.)

Zero-Phonon Lines

and Spectral Hole Burning in Spectroscopy and Photochemistry

With 64 Figures

Springer-Verlag Berlin Heidelberg New York
London Paris Tokyo

Dr. Olev Sild
Dr. Kristjan Haller
Institute of Physics, Estonian SSR Academy of Sciences
SU-202400 Tartu, Estonian SSR, USSR

Revised and updated translation of the Russian edition:
Bez fononnie linii v spectroscopii i fotokhimii © Izd. AN Est.SSR, Tartu 1986

ISBN 3-540-19214-X Springer-Verlag Berlin Heidelberg New York
ISBN 0-387-19214-X Springer-Verlag New York Berlin Heidelberg

Printing: Weihert-Druck GmbH, D-6100 Darmstadt
Binding: J. Schäffer GmbH & Co. KG., D-6718 Grünstadt
2157/3150-543210

Foreword

Lasers are playing a more and more dominant role in modern optical spectroscopy, offering an increased potential for high resolution and, thus, for more detailed spectroscopic information on dynamic and structural parameters. This book on *Zero-Phonon Lines and Spectral Hole Burning in Spectroscopy and Photochemistry* gives a concise and very useful survey of some of the pioneering and current work on solid state spectroscopy of various groups in the USSR. It focusses on the optical Mössbauer analogue, the "zero-phonon line", and "hole burning" spectroscopy, a method which increases the resolution well beyond the zero-phonon linewidth. In this context, the present work is complementary to *Persistent Spectral Hole-Burning: Science and Applications* (ed. by W.E. Moerner, Springer, Berlin, Heidelberg 1988), which deals in more detail with the various aspects of laser spectroscopy with ultrahigh spectral resolution. Zero-phonon lines and an understanding of the various phonon coupling mechanisms which are treated in this book are a prerequisite for applying and understanding techniques of ultrahigh resolution such as hole-burning or optical echoes.

Bayreuth, March 1988 *D. Haarer*

Preface

The investigation of zero-phonon lines (ZPLs) is one of the foremost and most informative fields of present-day condensed-matter spectroscopy. Along with its definite function in physical cognition and investigation methods, the spectroscopy of ZPLs is also gaining purely practical applications. This is due to the fact that ZPLs are extra-sensitive quantum-mechanical detectors. There is no doubt that this rapidly developing branch of spectroscopy will acquire a still greater academic as well as applied importance in the near future.

The investigation of ZPLs started in the first half of the 1950s with the theoretical treatment of the optical spectra of impurity crystals. Somewhat later, in 1958, the Mössbauer effect was discovered. It soon became clear that an optical ZPL and a Mössbauer (zero-phonon) line are, in principle, each other's analogues (some historical aspects of crucial events in this field of investigation are considered in the first contribution of this volume). In the case of a ZPL the spectrum is connected with electron transitions, in the latter case with nuclear transitions in the impurity centre of a crystal. In both cases, the ZPL of the spectrum is extremely narrow with a high Q-factor, its frequency and shape being extremely sensitive to (and hence informative about) the structure and parameters of the impurity centre, the processes going on in the centre and the physical properties of the surrounding medium.

A real crystal, however, is always inhomogeneous. The changes of the physical properties of the environment change the frequencies of impurity ZPLs, which results in so-called inhomogeneous broadening. As an electron transition associated with an impurity centre is much more sensitive to changes in the environment than a transition taking place in an impurity nucleus, the inhomogeneous broadening of the Mössbauer line is of the order of the linewidth itself, while that of an optical ZPL may exceed the homogeneous linewidth millions of times.

ZPL spectroscopy made a flying start with the advent and development of laser technology. The investigation of homogeneous optical spectral lines became feasible with the help of highly selective laser excitation, which enables centres with a resonance frequency to be distinguished from an inhomogeneous ensemble of impurity centres.

The role of selective laser excitation in the physics of ZPLs turned out to be decisive in two aspects. Firstly, it eliminates the disturbing influence of inhomogeneity and, secondly, it initiates irreversible processes in the centres of frequency resonance, which result in the "burning of the hole" into the inhomogeneous spectral line. Besides the possibility of obtaining information about homogeneous spectral lines, hole-burning spectroscopy offers excellent application prospects (information storage, narrow-band optical filters, etc.). Thanks to the advancement of hole burning we also have such an exciting and educational field of spectroscopy as space-and-time-domain holography.

The present volume is intended to give a selection of current results of experimental and theoretical investigations as well as survey-articles on the study of ZPLs in impurity centres, glasses, large organic molecules, and *in vivo* biosystems. Some applied aspects of the physics of ZPLs are also treated, primarily those of spectral hole burning and space-and-time-domain holography.

The authors of this volume are respected scientists working on ZPL investigations, a number of them having made major contributions to the field. Most of the investigations reviewed have been performed at the Institute of Physics of the Estonian SSR Academy of Sciences, which since the 1960s has been one of the leading centres in the field. The leader of the investigations in this institution is Academician Karl Rebane.

The spectroscopy of zero-phonon lines is a realm where both the academic and practical interests of physicists, chemists and biologists are intertwined. Therefore, the present book should be of interest to a wide variety of readers from students of relevant topics to top specialists in the field.

Thanks are due to Mrs. E. Vaik for revising the manuscript and to Miss L. Pedosar and Miss L. Juhansoo for typing it.

Tartu, December 1987

O. Sild
K. Haller

Contents

Zero-Phonon Lines in the Spectroscopy and Photochemistry of Impurity-Doped Solid Matter

K.K. Rebane

Presidium of the Estonian SSR Acad. Sci.
SU-200103 Tallinn, Estonian SSR, USSR

Abstract

A brief account of zero-phonon lines in the low-temperature spectra of impurity atoms and molecules in solid matrices - crystals, glasses, polymers - is given. Inhomogeneous broadening, its elimination and applications by photoburning of spectral holes are considered. Some remarks on the history and future of the topics are made. A variety of physicists, chemists, biologists, applied spectroscopists and also data processing specialists are considered as potential readers.

1. Introduction

The zero-phonon line (ZPL) in the optical spectra of the atoms and molecules introduced as impurities in solid matrices is a beautiful phenomenon of the physics of molecules and crystals, interesting by itself and useful for a number of applications.

At low temperatures ZPL, especially that of the purely electronic transition - the purely electronic line (PEL), are narrow and with a very high peak intensity [1, 2].

These unique features of the spectra of crystals and large molecules serve as the corner stone for high-resolution spectroscopic studies and spectrally selective technological applications of low-temperature impurity crystals. The photoburning of persistent spectral holes (SHB) in the inhomogeneously-broadened bands of ZPL [3,4] provides a new basic tool: the illumination-assisted control of the absorption coefficient and the refractive index of the solid matter activated by impurities.

Among the present and possible future applications are: wide-aperture absorption filters of high spectral selectivity; the measurements of emission spectra with the hole-burning medium serving as a spectrometer; optical spectral memory for computers [5,6]; time-and-space-domain holography of ultrafast events of pico- and nanosecond duration [7]; data storage and data processing by means of this new version of holography (see also review papers in [8]).

A special feature of SHB should be pointed out. Many of the phenomena and methods of laser optics may be considered to be to some extent analogs of the earlier achievements of radiospectroscopy. It is easy to understand because the latter had, from the very beginning, powerful sources of coherent radiation and started to deal with the effects of coherence and nonlinearity years before optics did. The well-known reason for it is that because of its high frequencies the particle-like features of optical radiation are much more pronounced and it causes additional difficulties for the creation and studies of coherence phenomena. Now SHB represents an optical phenomenon where high frequency of the hole-burning radiation becomes an essential advantage: it provides each photon with an amount of energy sufficient to initiate an electronic transition and guarantees an efficient hole burning. The radio frequencies are much less convenient (if not entirely impotent) to create persistent changes in the matter. SHB does not need high intensities of the exciting field. The dose of the irradiation is what matters. It may be collected by choosing proper exposition times. All experiments considered below have been performed with really modest intensities of laser illumination, so that the light-matter interaction was well within the limits of linearity. The doses depend on the quantum efficiency of SHB. The situation reminds us of colour photography, but SHB is much more selective both spectrally and spatially. The photosensitivity is considerably lower than that of photoemulsions.

The nature of SHB turns out to be especially useful in data storage and data processing: the beam of light carrying information (e.g. optical image of an event) performs with very high precision the space-and-frequency-dependent SHB. (Space actually means the coordinates at the surface of a plate of SHB substance.) The beam itself stores its spatial and frequency properties in the persistent long-living distribution of SHB profiles.

* * *

Optical ZPL are being studied in a number of laboratories in our country and abroad. The topic is expanding and includes now also problem of the molecular structure and energy transfer of systems interesting for chemistry and biology as well.

The corresponding spectroscopic methods have been developed in the Institute of Spectroscopy of the USSR Academy of Sciences [9,3] and in the Institute of Physics at Tartu [9] (see also the pioneering paper by SZABO [10] on fluorescence line narrowing and review papers [8,11-14]). By now a number of well-equipped laboratories abroad have entered the field and hole-burning spectroscopy is developing rapidly. It should be pointed out that one of the most exciting items - the hole-burning recording of the time behaviour of picosecond light pulses and the space-and-time-domain holography of ultrafast events - has been performed up to now only in the Institute of Physics [7,15].

Recalling the history, we certainly have to mention the pioneering work by E. Shpol'skii who invented the method of producing line spectra (quasi-line spectra) of large molecules in solid matrices [16]. The Shpol'skii effect is a considerable narrowing of the inhomogeneous broadening of spectral bands by means of crystal chemistry - by choosing for certain impurity molecules (mostly aromatic) appropriate solvents (mostly paraffins) to act as solid matrices. For that selected body of systems (the Shpol'skii systems) it became possible to observe spectral lines which at that time were surprisingly narrow for large molecules - with the width up to 1 cm^{-1}. But, on the other hand, the theoretically-predicted limit linewidth at $T = 0$, i.e. the radiative width $\Gamma_0(0)$, determined by the radiative lifetime τ_e of the excited electronic state, was still some three or four orders of magnitude narrower [1]. Indeed, for allowed transitions with the lifetime $\tau_e = 10^{-7}$ s at liquid helium temperatures the radiative linewidth $\Gamma_0(T) \approx 10^{-4}$ cm^{-1}, being still less for forbidden transitions with longer lifetimes of the excited electronic state.

The residual inhomogeneous broadening of ZPL in Shpol'skii spectra can be eliminated by making use of modern laser spectroscopy methods: the frequency-selective excitation of fluorescence (also fluorescence line-narrowing (FLN) and site-selective spectroscopy) [10,9] and the photoburning of spectral holes (SHB) [3,4,11,8]. These methods, especially SHB, enable one to get experimentally the linewidths close to the theoretical limit values and extrapolate them for $T \to 0$ quite precisely to the natural linewidths, which in the absence of radiationless processes means $\Gamma_0(0)$. It is also important that these methods allow one to overcome the rather strict restrictions put by crystal chemistry on the formation of the Shpol'skii systems. Owing to new methods it is now possible to measure with high spectral resolution, $\omega/\Gamma_0(T) \approx$ $\approx 10^6 - 10^9$ (ω - transition frequency), almost all molecules frozen as impurities in rather arbitrary solvents. The system may be polycrystalline, glassy or polymeric and the concentration of impurities may be very low. An aspect important for the practice of spectral analysis arises: the determination of even very small impurity concentrations (e.g. carcinogenic 3,4-benzopyrene in gasoline) requires no special chemical preparation of the sample, it suffices to freeze and cool a natural mixture (e.g. gasoline delivered at gas stations) down to low temperatures. Naturally, if a really high sensitivity is desired (10^{-10} grams per gram and higher), cooling down to liquid helium temperatures and high-resolution spectral apparatus is needed.

Most briefly the main features of ZPL can be expressed by the statement that they are an optical analog of Mössbauer lines. This name was given by GROSS [17] based on the theoretical paper by TRIFONOV [18]. From the point of view of physics this is a good definition. In the case of optical ZPL the interaction of the electronic transition with phonons takes place due to the perturbation of the *coordinates* of vibrating atoms in the crystal (in an electronic transition the *potential* energy of lattice vibrations undergoes an abrupt change), while in the γ-transition in nucleus the *velocities* (momenta) of atoms are perturbed (the recoil momentum of the γ-photon alters the *kinetic* energy of vibrations). The vibrational structure of the spectra is in both cases determined by the Franck-Condon principle [1]. In a good approximation the vibrations of atoms in the crystal can be described as the vibrations of a manifold of harmonic oscillators. In this case the Hamiltonian of the harmonic oscillator is symmetric with respect to the interchange of the coordinate and the momentum, i.e. the dependence of the energy of a harmonic oscillator on both the momentum and the coordinate is the same - quadratic. This symmetry is the reason of the deep analogy between optical ZPL and Mössbauer zero-phonon lines.

On the other hand, the term "optical analog of the Mössbauer line" is not justified from the historical point of view: the main features of optical ZPL were established already in a theoretical work by KRIVOGLAZ and PEKAR [19] five years before Mössbauer's discovery (and also earlier than the result by DICKE [20] which established that the spectral lines of a gas enclosed in a volume have a part free from the Doppler broadening).

In [19], a theory of impurity spectra is presented taking into account the dispersion of phonon frequencies. A formula is given describing the partial spectrum, i.e. the contribution to the spectrum by the vibronic transitions accompanied by the creation and annihilation of certain numbers of phonons. All partial spectra, with a single exception, have considerable widths caused by the widths of phonon frequency bands. The exceptional partial spectrum corresponds to the purely electronic transition, i.e. the transition not accompanied by any change of phonon population numbers (zero-phonon transition). This partial spectrum is described in [19] by a δ-function, i.e. it has a zero width, although the interaction with vibrations is not considered to be samll. The comment on ZPL is really short - it is stated that "The corresponding spectrum is not a band but a line" (eight words in Russian; see in [19] below formula (126)). The little attention to ZPL can be understood if we remember that at those early times the theory was concerned with the description of the *broadening* of spectral lines into broad bands as a result of introducing atoms into solid matrices, while the existence of rather narrow lines in the spectrum of the same impurity atom was considered to be a natural feature - they were well known for atoms in the gas phase. No attention was paid to the most important point of the phenomenon: in the gas phase spectral lines are broadened by the Doppler effect, while in the solid matrices they are not. Further, combining formulae (126) and (125) in [19], one can see that the intergated intensity of ZPL is proportional to a coefficient now known as the Debye-Waller factor. As a result, two main features of ZPL are actually present in [19]. The proper wording of the result, however, is actually missing and it was difficult to "dig out" the information. The possibility of measuring extraordinary narrow ZPL in the impurity spectra of solids was left without attention for almost a decade. In experiments,only inhomogeneously-broadened lines were being observed and there was no search to eliminate inhomogeneity for ZPL.

Quite different was the story of the resonance γ-lines discovered by R.Mössbauer in 1958 and named after him. This work attracted immediate attention and brought the author the Nobel prize in 1961. As it was already mentioned, Mössbauer's line is a typical zero-phonon line which could be referred to as a γ-radiation analog of an optical purely electronic ZPL. But Mössbauer found and proved his effect experimentally, which is always most important in physics. Another important advantage was that there is no *inhomogeneous* broadening in γ-spectra, what was the reason why the unique features of the homogeneous Mössbauer line showed up almost in the first experiments. Further, probably the γ-spectroscopists were well prepared to accept Mössbauer's discovery: the minds were tuned to "resonance" and the apparatuses needed to start the Mössbauer experiments were readily available.

In 1960 the first laser appeared. High-resolution optical spectroscopy got a new powerful tool. Interest towards activated impurity crystals increased in search for new lasing media. Pronounced ZPL (of course, along with inhomogeneous broadening but not more than some 1 cm^{-1}) were observed in the spectra of rare-earth impurity ions [21] and also in ruby [22]. Shpol'skii spectra attracted considerable attention but the theoretical work [19] was so far out of look that no adequate theory was considered to clarify *homogeneous* ZPL in the Shpol'skii effect. The paper by Krivoglaz and Pekar was cited but not in connection with the theory of purely electronic ZPL.

An important contribution was made by TRIFONOV [18] who showed an amazing analogy between the Mössbauer line and optical ZPL. An experimental proof of the characteristic temperature dependence was presented by GROSS et al. on an example of localized excitons in CdS crystal [17].

In [23], a quantitative theory of the Shpol'skii effect was elaborated and also its analogy with the Mössbauer effect demonstrated, taking into account also local vibrations. Further, the analog with the Mössbauer effect was considered in a general case - by omitting the assumption of harmonicity of lattice vibrations. It was demonstrated both qualitatively via Franck-Condon principle and mathematically through Fourier transform theorems [24,1].

From the point of view of the present knowledge it is important that in the theory [23](see also [1])an adequate estimation of the role of the *inhomogeneous broadening* was presented. It was pointed out that the genuine homogeneous purely electronic ZPL can be many orders of magnitude narrower than those appearing in Shpol'skii spectra. In [1], the possibilities of overcoming the inhomogeneous broadening were also discussed.

The theory [23,18,1] was welcomed by the molecular matrix spectroscopy community and, in particular, by E.V.Shpol'skii.

At that time there really were rather good grounds to believe that the theory is consistent. Nevertheless, there was no experimental evidence of several important points of the theory for *molecular* systems. The experimental situation was much better for impurity ions in a single crystal, in particular, for the spectra of ruby and rare-earth ions in ionic crystals: the ZPL in low-temperature spectra were narrow (1 cm^{-1} and less) and it was understood that this value was probably determined by the inhomogeneous broadening and that the homogeneous ZPL might be still much more narrow. (Concerning ruby a consistent point of view was expressed already in [22].)

In the case of large impurity molecules, however, including Shpol'skii systems, there were some doubts which even increased during 1964 - 1970. These doubts were initiated, first of all, by the circumstance that nobody had detected phonon sidebands. Secondly, the narrowest experimentally-observed lines were still some 3 - 4 orders of magnitude broader than the optical analog of the Mössbauer line should have been. It was generally understood that the latter circumstance was due to inhomogeneous broadening. It turned out that the former was also caused by the influence of inhomogeneity.

Thus, by the end of the sixties the theory of the Shpol'skii effect was put under question. On the other hand, by that time at the Institute of Physics in Tartu better experimental facilities became available to study the spectra of small impurity molecules. One decided to cross the line from the pure theory to experimental studies also in the problem of phonon sidebands in the Shpol'skii effect. It must be underlined that this decision was taken only after serious considerations of the situation and it is only now that it seems quite natural that in the Institute of Physics there are also experiments with large impurity molecules in solid matrices. Phonon sidebands were first detected in both luminescence and excitation spectra of perylene molecules in n-hexane [25]. As a result, for the first time a characteristic picture of a ZPL (still inhomogeneously broadened) accompanied by a phonon sideband was observed for a high-molecular system. An important point here was the application of *selective excitation* to a certain component of the Shpol'skii multiplet, i.e. the elimination of the manifestation of the inhomogeneous structure of the matrix,which expressed itself as "splitting" of the ZPL into a multiplet. In the case of a broad-band excitation the emission spectral lines belonging to different components of the multiplet are superimposed, which most severely mixes up different ZPL and their phonon sidebands. As a result the actual structure is lost.

An important role in confirming and gaining popularity for the theory of ZPL and fine structure of vibronic spectra was played by the studies of small impurity molecules (O_2^-, S_2^-, NO_2^- and their analogs) in alkaline halide crystals (see [26], review [27]). Luminescence spectra, along with other components of secondary emission (hot luminescence, Raman scattering), are quite characteristic and informative in these systems because of clear-cut ZPL (still inhomogeneously broadened!) and structured phonon sidebands (PS).

Notable was also the process of establishing ZPL in the spectra of chlorophyll molecules. Some time ago it was not clear at all whether such a "specific" molecule as chlorophyll does possess ZPL and,in case it does, can they be as narrow as they are in the case of other large molecules. By now, mainly due to the work by R.Avarmaa, it is clear that chlorophyll-like molecules do have ZPL and that they serve well for high and very high resolution (in the latter case the inhomogeneous broadening is eliminated) matrix spectroscopy (see reviews [28]). ZPL have been identified also for chlorophyll in native systems [29].

* * *

The analogy between optical and Mössbauer ZPL helps to understand both of them better. It is quite useful in communicating the essence of optical ZPL to the people well aquainted with the Mössbauer effect and vice versa - explaining the Mössbauer ZPL to those who are on good terms with optical ZPL. As phonons are always present in solid systems and take part in the shaping of any spectra, the theory of ZPL makes up an important chapter in modern spectroscopy of solid systems.

The values of the experimental homogeneous widths of purely electronic ZPL at low temperatures, $\Gamma_0(T)$, lie now in the same interval as those of Mössbauer lines. But the spectral resolution $\omega/\Gamma_0(T)$ (Q-factor) of optical ZPL remains considerably less than that for Mössbauer measurements, owing to the high frequency of γ-transitions in the Mössbauer effect as compared to optical frequencies. In the case of optical transitions there are still further possibilities connected with long lifetimes of the excited states of forbidden transitions. Investigations in this direction belong to the future.

Figure 1 represents schematically a characteristic view and the temperature dependence of a homogeneous impurity spectrum with a ZPL and its phonon sideband. The basic interaction with phonons does not broaden ZPL but creates the phonon sideband well separated from the ZPL and also causes a rapid transfer of the integral intensity from ZPL to PS. The width of ZPL in the temperature region of liquid He is $\Gamma_0(T) = 10^{-3} - 10^{-7}\ cm^{-1}$, the ratio of the peak intensity I_0 of a ZPL to the peak intensity of its phonon sideband I_p (i.e. the intensity of ZPL to its own background), $I_0/I_p = 10^4 - 10^9$, and the spectral resolution, $\omega/\Gamma_0(T) = 10^7 - 10^{11}$. Temperature broadening, although very important for narrow ZPL, still represents a second-order effect with respect to the interaction of phonons with the electronic transition.

Optical ZPL possess a specific feature as compared to Mössbauer ZPL - an extremely large inhomogeneous broadening. As it was already mentioned, it is a seri-

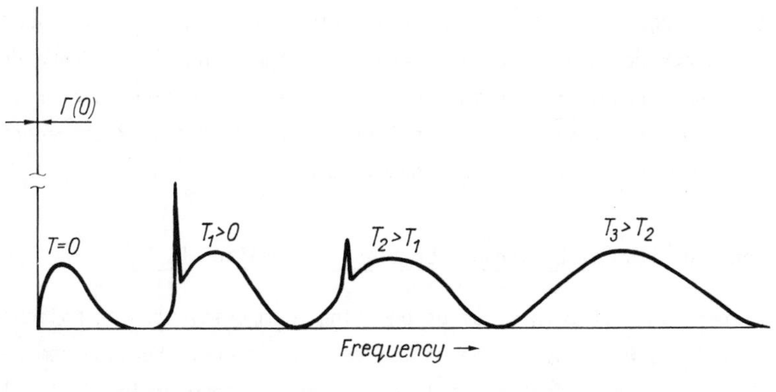

Figure 1: A purely electronic line, its phonon wing and their dependence on the temperature T in the homogeneous spectrum of impurity absorption

ous disadvantage in spectroscopy and it is one of the reasons why optical ZPL were not met enthusiastically from the very beginning. On the other hand, as we shall see further, owing to photoburning of spectral holes the inhomogeneous broadening turns into a useful feature of a number of applications.

Electronic transition is influenced by inhomogeneities in the structure of the matrix surrounding the impurity: ZPL frequencies are slightly different depending on the position and (or) orientation of the impurity in the matrix. The narrower the line is, the more severe is the damage to its quality (both the width and peak intensity) caused by inhomogeneity. The smallest differences of the order of 1 cm^{-1} in transition energies caused by strain and electric fields due to other impurity defects and dislocations mean already enormous inhomogeneous broadening in comparison with the homogeneous linewidth of 10^{-3} cm^{-1}. It can reach hundreds of cm^{-1}, i.e. exceeding the homogeneous width by hundreds of thousands times. So,due to their extreme narrowness and sensitivity ZPL easily become drowned in inhomogeneous broadening.

There is still another important aspect: there are really very many systems which possess narrow optical ZPL at low temperatures - thousands of molecules and other kinds of impurity defects in hundreds of various matrices, while there are only some tens of nuclei suitable for the Mössbauer effect. To the systems with distinct homogeneous ZPL one can account also rather big molecules, for example, chlorophyll, porphin and their derivatives, which are of considerable interest to molecular physics and biology.

* * *

Papers contributed to the present issue are concerned with the results obtained during recent years in the field of experimental and theoretical study of ZPL. Possible practical applications are only mentioned. The number and fields of the latter are increasing and they deserve at least a special review.

The present issue is meant not only for physicists in the field of ZPL but also for chemists, biologists, people in applied optics, holography, data processing. Therefore, it seems reasonable to give here a short account on the fundamentals of the theory of homogeneous and inhomogeneous ZPL spectra. References of review articles are given, which contain exhaustive lists of publications. Most recent publications are referred to in the contributions to the present issue.

2. Homogeneous Spectrum of Electron-Vibrational Transition [1,2]

Homogeneous spectrum is the spectrum of one single impurity atom or molecule. An ensemble of impurities will have, even at low concentrations, the same homogeneous spectrum only in an ideal case when all impurities are absolutely identical and undergo precisely identical changes under the influence of the surrounding matrix. This does not happen in reality and an impurity spectrum is always *inhomogeneous*. It is formed by a multitude of partially overlapping homogeneous spectra. So, inhomogeneous spectra are determined by the qualities of homogeneous spectra. The possibilities of extracting the homogeneous spectrum out of the inhomogeneous one depend greatly on the properties of the former.

2.1 Integral Intensity of PEL and Phonon Sideband, the Debye-Waller Factor; the Width of PEL and its Temperature Dependence. ZPL of Local Vibrations

Let us first consider an impurity centre with no local vibrations. An electron-vibrational (vibronic) transition between its two electronic states is accompanied by the creation and annihilation of phonons of the lattice modes of the matrix. Due to acoustic modes, whose frequencies start from zero, the spectrum of the transition always covers, in principle, a more or less broad interval of frequencies and is continuous. Nevertheless, the theory [1] tells us that the homogeneous spectra of the transition look like as shown in Figure 1. At $T = 0$ there is a very sharp and narrow ZPL with high peak intensity as compared to the peak intensity of phonon sideband. The width of ZPL, $\Gamma_0(0)$ (in cm^{-1}), is determined by the lifetime τ_1 of the excited electronic level (c - velocity of light)

$$\Gamma_0(0) = (2\pi c\tau_1)^{-1} . \qquad (1)$$

Here there is no Doppler broadening because impurity centres are bound to the macroscopic mass of the matrix. The reason is the same what makes the Mössbauer γ-resonance line free of Doppler broadening.

For allowed optical transitions in the absence of nonradiative quenching we have $\tau_1 \simeq 10^{-7} - 10^{-8}$ s and, correspondingly, $\Gamma_0(0) \simeq 10^{-4} - 10^{-3}$ cm^{-1}. For forbidden transitions ZPL can be still narrower.

At $T \neq 0$ ZPL is broadened as a result of scattering of phonons on impurities (or as a result of interaction with elementary excitations of any other nature, initiated by thermal motion). Scattering processes lead to abrupt changes of the phase of the wave function of the excited electronic state in the impurity (phase relaxation). The electron remains on the excited level but its wave function is altered. The latter means that the lifetime in that particular quantum state is shortened and the corresponding homogeneous spectral width is broadened as comaperd to the width determined by the population lifetime of the level (energy lifetime). Zero vibrations do not cause dephasing and so at $T = 0$ there is no additional broadening. The density of the phonons induced by thermal motion increases with temperature and this is the reason of the temperature broadening of ZPL.

It is reasonable to introduce two characteristic lifetimes - energy (also longitudinal) relaxation time, τ_1, and phase (also transverse) relaxation time, τ_2^*, (pure dephasing time). Energy relaxation also causes phase relaxation and the resulting excited electronic state's lifetime τ_2,which determines the homogeneous width of PEL, is the following:

$$\Gamma_o(T) = (2\pi c\tau_2)^{-1} = (\pi c)^{-1}[(2\tau_1)^{-1} + (\tau_2^*)^{-1}] . \tag{2}$$

If nonradiative damping processes are present, then $\tau_1^{-1} = \tau_{1opt}^{-1} + \tau_{1q}^{-1}$, where the first term corresponds to the rate of radiative processes and the second one, to the rate of nonradiative processes. Usually, the temperature dependence of τ_{1opt} can be neglected but τ_{1q} and τ_2^* depend essentially on T. At liquid helium temperatures nonradiative processes are usually absent ($\tau_{1q} = \infty$) or remain on their low-temperature limit of quantum tunnelling ($\tau_{1q} = \text{const}$) and the dephasing caused by phonons becomes the main reason for the temperature broadening of ZPL. To estimate the amount of temperature broadening at low temperatures for large molecules as impurities in single crystals and polycrystalline matrices one can suppose that the phase relaxation time is roughly equal to the energy relaxation time at temperatures 1.8 - 4.2 K. The contribution of dephasing to the ZPL width increases rapidly with temperature and becomes overwhelming. For the same molecule in glassy or polymeric matrices the contribution of dephasing exceeds several times the line broadening caused by the population decay of the excited state already at 1.8 K (see [11-14,30]).

ZPL is accompanied by a broad continuous band - the phonon sideband - which corresponds to the transitions in which phonons are created and annihilated. The detailed structure and width of the PS are quite individual and depend on the local lattice dynamics of the impurity centre (i.e. on the spectral density of the phonons of the matrix and on phonon perturbations in the vicinity of the impurity) and on the strength of the interaction of the electronic transition with lattice vibrations (i.e. on the degree of the changes of the local dynamics caused by the change of the electronic state). Many phonons can take part in the transition and the width of the PS may essentially exceed that of the spectrum of acoustic phonons. The spreading of the latter may nevertheless serve as a starting point for the estimation of the width Γ_p of PS: $\Gamma_p = 10 - 100\ \text{cm}^{-1}$. Thus, ZPL is indeed a very sharp spectral line: it is $10^4 - 10^7$ times narrower than PS.

Besides the linewidths the peak intensities are also important. To get an idea about them one has to know integrated intensities. The relative integrated intensity of ZPL at T = 0 is expressed by the Debye-Waller factor $\alpha(0)$ [1]:

$$\alpha(0) \equiv S_o/(S_o + S_p) \equiv \exp(-\bar{n}_{St}) , \tag{3}$$

where S_o and S_p are, respectively, integral intensities of ZPL an PS; $\bar{n}_{St} = \sum_{s=1}^{N} P_s/(\hbar\omega_s)$ is the sum of Stokes losses expressed in phonon numbers; P_s is the Stokes shift in energy for the s-th mode; $2P_s = m_s\omega_s^2 q_{so}^2$, where m_s is mass, ω_s is the frequency and q_{os}, the shift of the equilibrium position of the oscillator.

With the increase of temperature the Debye-Waller factor decreases (monotonically if there are no local vibrations) and rather rapidly at that. For the basic model the theory gives [1]:

$$\alpha(T) \equiv S_o(T)/[S_o(T) + S_p(T)] = \exp[-\sum_{s=1}^{N} (2\bar{n}_s + 1)P_s/(\hbar\omega_s)] , \tag{4}$$

where the number of phonons $\bar{l}_s$ in the lattice mode s at the temperature T is given by $2(\bar{l}_s + 1/2) = \coth(\hbar\omega_s/2kT)$. At high temperatures, $kT >> h\omega_s$, $\bar{l}_s$ increases linearly with temperature, that gives an exponential decay of $\alpha(T)$. The decay of the ZPL intensity is the faster the bigger are the dimensionless Stokes losses of the thermally excited modes at the temperature T. When $T \to 0$, then $\bar{l} \to 0$ and formula (4) turns into (3). But it should be kept in mind that acoustic phonon frequencies start right from zero and the lowest frequency (i.e. of the longest wavelength) vibrations are always thermally excited. There are two reasons, however, why the influence of low-frequency vibrations on vibronic spectra can be practically neglected. The first is the very low density of these vibrational states in a crystalline matrix, i.e. the number of matrix oscillations, which can be thermally excited at $T \to 0$, decreases rapidly with the temperature. Secondly, the general considerations of the theory of oscillations tell us that the influence of the transition localized in the electronic shell of the impurity on the very long-wavelength acoustic modes, can be negligibly small, i.e. $P_s(\omega_s) \to 0$ when $\omega_s \to 0$. In glassy matrices, the density of low-frequency excitations of the matrix (phonons, two-level systems, pseudolocal vibrations) is considerably higher. In particular, at low frequencies a large contribution comes from two-level systems whose density $\rho(\omega_{TL})$ is considered to be approximately constant, i.e. $\rho(\omega_{TL}) = \text{const}$. That is why the first reason does not work. Possibly it starts to show itself at considerably lower temperatures as compared to crystals. There are special aspects about the second reason as well: low frequency can be realized in a glassy system locally in the vicinity of the impurity. As a result $\alpha(T)$ and especially, the temperature dependence of the ZPL width can have and have, in the case of glassy matrices, interesting peculiarities [11-14,30].

If nonradiative transitions are absent, the sum of the integrated intensities of ZPL and PS represents the total intensity of the electronic transition (oscillator strength of the transition), i.e. the same quantity which determines the radiative lifetime τ_1. In the same limits in which τ_1 does not depend on temperature, the sum of the two quantities (each of which rather sharply depends on temperature) is temperature independent:

$$S_o(T) + S_p(T) = \text{const.} \qquad (5)$$

So formulae (3) and (5) tell us that in the considered approximation the temperature dependence of the integrated intensities results in "the lossless pumping" of the intensity from ZPL to PS.

For the largest organic molecules in solid matrices, including Shpol'skii systems [16], the limit temperature, above which ZPL is almost absent, lies near 20 - 30 K. So ZPL does not only broaden with temperature but also loses its integral intensity. For many impurity systems the latter is the main reason why ZPL are absent at higher temperatures.

Figure 1 shows schematically the dependence of the absorption spectrum on temperature. The decrease of the Debye-Waller factor and ZPL broadening are presented. Luminescence spectra have the same features. For the basic theoretical model it is a mirror image of the absorption spectrum [1].

Let us note that the same expression (4) describes the behaviour of the Mössbauer γ-resonance line if Stokes shifts are replaced by the recoil energy R_s transferred to a vibrational mode s in the process of absorption or emission of a γ-photon.

This is another interesting aspect to the analogy between ZPL and Mössbauer lines, which is finally determined by the symmetry of the Hamiltonian of harmonic oscillators with respect to the interchange of the coordinate and the momentum.

2.2 Relative Width (Q-Factor) of ZPL. Peak Intensity

For the Q-factor - the ratio of the ZPL frequency ω_0 to its homogeneous width $\Gamma_0(T)$ - in the spectra at LH temperatures we get an estimate $Q = \omega_0/\Gamma_0(T) = 10^8$ if we take $\omega_0 = 10^4\ \mathrm{cm}^{-1}$ and $\Gamma_0(T) = 10^{-4}\ \mathrm{cm}^{-1}$. The high value of Q is a good starting point for high-resolution spectroscopy and it turns ZPL into a sensitive probe of processes and interactions in molecules and solids. This is the third aspect of the analogy with the Mössbauer line. But the high value of Q is not enough. For a convenient detection the peak intensity must be well above the background.

The peak intensity I_0 is proportional to the square of the transition dipole momentum multiplied by the Debye-Waller factor $\alpha(T)$, and reciprocal to the linewidth $\Gamma_0(T)$. If at $T = 0$ the width is equal to the radiative width (no non-radiative transitions present), then both the integral intensity $S_0(0)$ and the width $\Gamma_0(0)$ increase proportionally to the probability of the transition. Thus, the peak intensity remains unchanged as the transition probability increases. Further, the peak intensity is proportional to $\alpha(T)$, the maximum value of which is 1, being reciprocal to $\Gamma_0(T)$, the lowest value of which is the radiative width $\Gamma_0(0)$. In the simplest but still reasonable version of the theory, I_0 of the homogeneous impurity spectra cannot be larger than the homogeneous spectra (i.e. free from the Doppler broadening) of the same atom (molecule) in gas phase in the absence of collisions. In both cases - molecules in a matrix or molecules in vapour - the actual task is to get rid of the inhomogeneous broadening. In the case of Mössbauer lines inhomogeneous broadening is simply absent. An advantage of the high-resolution spectroscopy of ZPL in comparison with atoms in vapour is that in a solid matrix at low temperatures the *changes* of the inhomogeneity of the system are many orders of magnitude slower than the *changes* of the velocities of atoms in gases. It provides plenty of time to study the inhomogeneous structure fixed for a really long time and opens new horizons for creating and applying methods to eliminate or utilize the inhomogeneous broadening.

Returning to the spectral background we have to take into account, first of all, the phonon sideband - it forms the inevitable background on which the ZPL have to be measured. To estimate the ratio of I_0 to the maximum of PS we have to know the values of the Debye-Waller factor and the ratio of the ZPL width to the width of the phonon sideband. For impurity ions, such as Tl^+ or F-centres in alkaline halides, $\bar{n}_{st}$ constitutes some 30 - 100 and, respectively, $\alpha(0)$ is from exp(-100) to exp(-30). That is why there is no ZPL in the absorption and luminescence spectra of impurities even at $T=0$. It should be mentioned here that the ZPL have been well detected in the Raman scattering spectra of the same impurity systems and they can also be found in other nonlinear spectra, however, the integrated intensities of the ZPL in these spectra might be determined by combinations of the Franck-Condon factors different from those for the Debye-Waller factor in absorption.

Electronic transitions in inner electronic shells of rare-earth impurity ions are very well screened from the lattice and, therefore, generate almost no phonons. That is why $\bar{n}_{st} \ll 1$ and $\alpha(T)$ is close to 1. In this case the phonon sidebands are absent at low temperatures and often not detectable even at high temperatures because the temperature broadening mixes ZPL up with their phonon sidebands.

In Figure 1, the case of a medium interaction is presented. It corresponds to the parameters characteristic of organic molecules in organic matrices if one takes $T_1 = 4$ K, $T_2 = 15$K, $T_3 = 40$ K. The ZPL widths in Figure 1 are strongly exaggerated.

For estimates it is reasonable to suppose that $S_0(0) = S_p(0)$, which gives $\alpha(0) = 0.5$. Further we can suppose simple relations $S_0 = \Gamma_0 I_0$ and $S_p = \Gamma_p I_p$ (I_p is the maximum intensity of PS). For the ratio r of peak intensities we get then

$$r = I_0/I_p = \Gamma_p/\Gamma_0 \quad . \tag{6}$$

If we take $\Gamma_p = 100$ cm^{-1}, $\Gamma_0 = 10^{-4}$ cm^{-1}, we arrive at $r = 10^6$, which is really a good value for a signal to background ratio.

It should be noted that in time-gated spectra (time-dependent spectra, also transient spectra) the ZPL can be observed which are narrower than the radiative width [31,32].

2.3 Local Vibrations

If there are local vibrations present (and in the case of molecular impurities they always are), then in the limits of the basic model [1] every local vibration gives rise to a series of replicas of ZPL and phonon sidebands shifted with respect to the purely electronic frequency ω_0 by the frequencies $n\Omega_l$, where Ω_l is the frequency of the l-th local vibration ($n = \pm 1, \pm 2, \ldots$). Integrated intensities of the replicas are determined by Franck-Condon factors and by population of the initial vibrational levels corresponding to the vibronic transition under consideration. For high-resolution spectroscopy it is important that vibronic ZPL have homogeneous width broad as compared to purely electronic ZPL.

This results from the obvious reason that in the case of a transition involving local vibrations at least one of the energetic levels belongs to a vibrationally-excited state. The narrowest width of the latter is determined by its decay (longitudinal) lifetime τ_{1v}, which at T = 0 is about 10^{-11} s, that is for $10^3 - 10^4$ times shorter than the radiative lifetime of an electronic state. The homogeneous widths of vibronic ZPL are correspondingly larger, $\Gamma_v(0) \simeq 0.1 - 1.0$ cm^{-1}. $\Gamma_v(T)$ increases with temperature, but at low T, slower than $\Gamma_0(T)$, because the share of dephasing to the broadening, which gives the strongly temperature-dependent contribution to the linewidth, is approximately the same as in the case of the purely electronic line, and it results in a relatively small dephasing part in the linewidth at liquid helium temperatures.

Although vibronic ZPL are considerably broader and of lower peak intensity than PEL, they give rise to rather sharp peaks in the spectra. Often these peaks are superimposed with phonon sidebands and this situation has to be taken account of when interpreting the experimental data. For example, these peaks give rise to a new phenomenon - pseudosplitting of lines in the luminescence spectra created from *inhomogeneous* spectra by selective excitation [33] if the frequency of the local vibration is less than the inhomogeneous broadening.

2.4 On Inhomogeneous Spectra

Up to now we have dealt with homogeneous spectra. To obtain them the spectrum of a single molecule has to be measured or an impurity system has to be created where all impurities are embedded exactly in equivalent positions with respect to the matrix so that the differences of the energy of purely electronic transitions fall well within a 10^{-4} cm^{-1} interval. The first is a really hard experimental task, the second one being unrealistic (for solid matrices). In practice we always have to deal with inhomogeneous spectra of impurities.

In a good approximation one can take that the influence of the inhomogeneity results in a frequency shift of the homogeneous spectrum as a whole without any changes in its shape. In this case it is convenient to introduce a one-dimensional inhomogeneous distribution function (IDF), $\rho(\omega)$. $\rho(\omega)\Delta\omega$ gives for an electronic transition the fraction of impurities which have their purely electronic ZPL in the frequency interval $\Delta\omega$ around ω [34,35]. There is a great variety of IDF. For instance, in the case of a Shpol'skii multiplet IDF comprises several sharp peaks corresponding to the components of the multiplet, and a weak broad continuous background around them. Nevertheless, it is often sufficient to consider as a model a bell-shaped IDF with a single maximum.

Let us note that the inhomogeneously-broadened absorption (or other) spectra do not coincide with IDF: there are also absorption sidebands. Further, from a given IDF very different inhomogeneous and, especially homogeneous, spectra can be created (see review paper [35]).

An example of an inhomogeneous low-temperature absorption spectrum is presented in Figure 2. This part of absorption which is realized through ZPL has a hidden fine

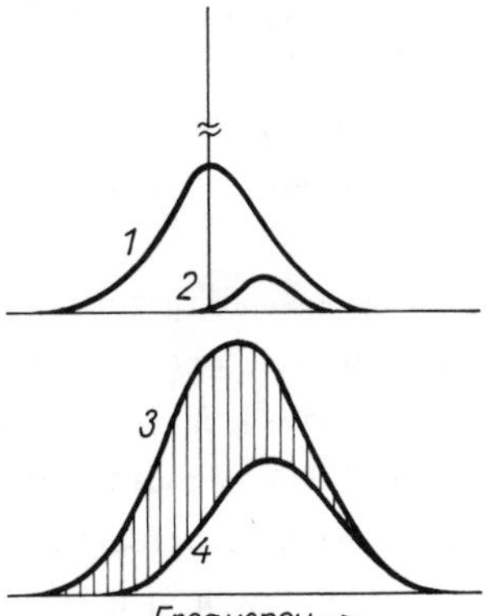

Figure 2: 1 - The inhomogeneous distribution function; 2 - the homogeneous spectrum of impurity absorption; 3 - the inhomogeneous spectrum of the absorption consisting of two parts: a continuous absorption due to phonon wings (4) and the absorption on the ZPL (between curves 3 and 4), possessing a latent selectivity. The figure is depicted for a Debye-Waller factor $\alpha(0) = 0.5$

structure that forms the basis for the high spectral sensitivity of impurity systems. High values of the peak intensities of homogeneous ZPL mean very large absorption cross-section for the radiation of the ZPL frequency ω_0. If the exciting light is nearly monochromatic, the transitions in other impurities, with frequencies shifted more than the sum of the excitation linewidth and the homogeneous width of ZPL, are out of resonance and absorb considerably less.

The application of this remarkable feature serves as the bases for the methods of site-selective spectroscopy and photoburning of spectral holes as well as their variations. Absorption in phonon sidebands is truly a continuous one and determines a continuous background absorption.

It is interesting to recall how the role of inhomogeneous broadening has progressed. Initially it was negative: inhomogeneous broadening was the reason why optical purely electronic ZPL could not serve for really high-resolution spectroscopy. Then laser methods of eliminating inhomogeneous broadening by means of selective excitation of luminescence and photoburning of spectral holes introduced high-resolution ZPL spectroscopy. Now, hole burning (HB) as a method of the control of optical properties of the matter by illumination has turned inhomogeneous broadening into a *useful* feature: the larger it is the wider is the spectral band available for HB applications. Broad inhomogeneous bands are especially useful for information storage and processing.

3. Conclusions

Zero-phonon line is a beautiful feature of impurity spectra in solids. Sometime it is mentioned as a unique property. If one refers to fine spectral characteristics, then it is true. On the other hand, ZPL are present in the low-temperature spectra of thousands of large and small molecules, atoms and ions imbedded as impurities in hundreds of ordered and disordered organic and inorganic matrices. From that point of view ZPL is not just a nice feature of only a single or a few unique systems, but a rather widespread phenomenon. It means, firstly, that there is a large body of systems subject to high-resolution ZPL spectroscopy and photochemistry and, secondly, it opens good outlooks for finding the systems well matching the requirements of various technological applications.

To have well-pronounced ZPL the liquid helium temperatures are obligatory. Nowadays laser studies at temperatures 2 - 20 K are quite a conventional method, available also for industrial applications. On the other hand, it is possible that there are systems having, on the one hand, rather distinct homogeneous ZPL at liquid nitrogen temperatures and even at room temperature, and, on the other hand, considerable inhomogeneous broadening. Here we are talking about search for really unique representatives among the multitude of ZPL systems.

At present three directions of ZPL studies may be mentioned. Firstly, the continuation of the theory and experiment of ZPL. Here belong the investigations of ZPL in disordered matrices (especially at sub-Kelvin temperatures), in time-dependent secondary emission spectra (theory predicts that the observed spectral lines can be narrower than the natural width; up to now experiments have been performed only for Mössbauer lines [36]), in nonlinear optical spectra, in the conditions of energy transfer between impurities (especially systems of biological interest), at high temperatures (liquid nitrogen and room temperatures), in liquids (including quantum fluids), and also ZPL of strongly forbidden transitions. Secondly, ZPL-based studies of the energy levels,energy transfer and transformation of solids and molecules, within the rapidly-developing field of conventional matrix spectroscopy and matrix photochemistry of molecules the studies of model and native biological molecular systems, biologically important large molecules, structure and processes in glasses and other disordered organic systems, and photochemical reactions of high spectral selectivity. Thirdly, scientific and practical applications of the possibilities to perform, by means of selective bleaching, light-controlled manipulations of absorption and the refractive index of solid media. To this may be attributed the manufacturing of super-narrow-band wide-aperture spectral filters, holography, including space-and-time-domain holography [7], selective optical memories for computers [6,8,37].

A few remarks on two of the present applications should be added.

Time-and-space-domain holography [7,15] is the most exciting and educational application of photoburning of spectral holes (see also [39]).

The ideas of the time-and-space-domain SHB holography have many beautiful applications and not only in image or event storage and processing. They may also be useful in the studies of solids by means of the new holography. For instance, the time-of-flight experiments may be arranged in such a way that before entering the sample under study the picosecond pulse is divided into two parts, one of which goes through the sample, the other is guided around it and time-delayed in such a way that it hits the HB recording medium behind the sample at the same spot as the first part, creating an information-rich interference pattern in the spectral domain (if the beams are parallel) or the spectral-space domain (if they hit the pellet at a nonzero angle). The pattern is stored by the HB pellet and may be thoroughly studied both spectrally and spacially or by applying various probing pulses. Actually it would be a new version of high-spectral-resolution holographic studies of the properties of the matter, which might open new horizons both for the investigations of energy levels and the processes in molecules and solids.

One of the promising applications is HB photoelasticity [38]. The idea is simple: strains shift and broaden holes, which results in the changes of transparency. The latter may be correlated to the strains by calibration measurements.

The advantages in comparison with the conventional, i.e. polarization-plane rotation, photoelasticity are the following:

1. High sensitivity. Very narrow holes created by narrow-line lasers in proper systems are well able to detect (and store!) even the changes in strains caused by the daily variations of the athmospheric pressure. The study of strains in very thin samples (shells) becomes also available.
2. New possibilities of storing the picture of strains are opened: the spectral hole can be burnt when the forces are applied to the sample and the changes in transparency studied *after* the forces are taken off and the strains released. It enables us to burn a spectral hole at some particular frequency for a particular situation of forces applied to the sample, another one at another frequency, fixing a different situation, and so on. In this way we can store a number of strain fields and study all of them after the sample is released.

If holes are burnt fast enough, then such sequence of spectral holes can store the time development of the strain field, e.g. vibrations of the sample. Currently the shortest hole burning time is 30 ns, but the times even $10^3 - 10^6$ times longer can still reveal a lot about quite fast dynamic processes.

On the other hand, liquid helium temperatures are required for HB and for detecting hole shapes. The storage temperature may be higher, sometimes even room temperatures are acceptable (see [37]). Once more we can see how important it is to find systems with narrow homogeneous ZPL which form a broad inhomogeneous band. Maybe the impurity states in multilayer quantum wells can serve as candidates to build such systems.

* * *

It should be pointed out that actually HB stores only the spectral doses of irradiation (closely correlated to intensities if the latter are not too strong). No straightforward storage of phases takes place. It holds also for holography: even in the case of two pulses separated in time (reference and signal pulses) actually only the doses are stored. The real problem is how to construct the whole pulse in such a way that the intracorrelations (stored via doses again!) respond to properly-chosen reading pulses in a proper way to extract the desired information. The simplest possibility is to divide the whole pulse into a δ-pulse and signal pulse, but it is not the only one. Naturally, what matters is the phase relaxation time of the excited electronic state of the impurity. This has to be long in comparison with the whole pulse duration, otherwise some essential information will get lost. The relation between the stored doses and intensity distribution in the signal matters as well, but it is rather a problem of experimental technique and theoretical interpretation.

References

1 K.K. Rebane: Elementarnaya teoriya kolebatel'noj struktury spektrov primesnykh tsentrov kristallov (Nauka, Moskva 1968) [English transl.: Impurity Spectra of Solids (Plenum Press, New York 1970)] and references therein

2 A.A. Maradudin: Rev. Mod. Phys. 36, 417 (1964); A.A. Maradudin, E.W. Montroll, G.H. Weiss: Theory of Lattice Dynamics in the Harmonic Approximation (Academic Press, New York 1963); R.H. Silsbee, D.B. Fitchen: Rev. Mod. Phys. 36, 433 (1964); N.N. Kristoffel: Teoriya primesnykh tsentrov malykh radiusov v ionnykh krystallakh (Nauka, Moskva 1974); I.S. Osad'ko: in Spectroscopy and Excitation Dynamics of Condensed Molecular Systems, ed. by V.M. Agranovich, R.M. Hochstrasser (North-Holland, Amsterdam 1983) p. 437; M.N. Sapozhnikov, V.I. Alekseev: Phys. Status Solidi b120, 435 (1983); A.M. Stoneham: Theory of Defects in Solids, Vol. 1,2 (Clarendon Press, Oxford 1975); D. Hsu, J. Skinner: J. Chem. Phys. 83, 2097 (1985); J.L. Skinner, D. Hsu: Adv. Chem. Phys. 65, 1 (1986) and references therein

3 B.M. Kharlamov, R.I. Personov, L.A. Bykovskaya: Opt. Commun. 12, 191 (1974)

4 A.A. Gorokhovskii, R.K. Kaarli, L.A. Rebane: Pis'ma Zh. Eksp. Teor. Fiz. 20, 474 (1974) [English transl.: JETP Lett. 20, 216 (1974)]; Opt. Commun. 16, 282 (1976)

5 G. Castro, D. Haarer, R.M. Macfarlane, H.D. Trommsdorf: US Patent 4,101,976 (1978); D.M. Burland: US Patent 4,158,890 (1979); A. Szabo: Proceedings of the International Conference on Lasers'80, New Orleans, Lousiana, Dec. 15-19, 1980, ed. by C. Collins (STS Press. McLean 1981) p. 374; K.K. Rebane: Proceedings of the International Conference on Lasers'82, New Orleans, Lousiana, Dec. 13-17, 1982 (STS Press, McLean 1982) p. 340

6 M. Romagnoli, W.E. Moerner, F.M. Schellenberg, M.D. Levenson, G.C. Bjorklund: J. Opt. Soc. Am. B1, 341 (1984); F.M. Schellenberg, W. Lenth, G.C. Bjorklund: Appl. Opt. 25, 3207 (1986); W. Lenth, R.M. Macfarlane, W.E. Moerner, F.M. Schellenberg, R.M. Shelby, G.C. Bjorklund: Proc. SPIE 695, 216 (1986)

7 P. Saari, R. Kaarli, A. Rebane: J. Opt. Soc. Am. B33, 527 (1986)

8 W.E. Moerner (ed.): Persistent Spectral Hole Burning. Science and Applications, (Springer) in press

9 R.I. Personov, E.I. Al'shits, L.A. Bykovskaya: Pis'ma Zh. Eksp. Teor. Fiz. 15, 609 (1972) [English transl.: JETP Lett. 15, 431 (1972)]; Opt. Commun. 6, 169 (1972)

10 A. Szabo: Phys. Rev. Lett. 25, 924 (1970)

11 L.A. Rebane, A.A. Gorokhovskii, J.V. Kikas: Appl. Phys. B29, 235 (1982); L.A. Rebane: Zh. Prikl. Spektrosk. 34, 1023 (1981) [English transl.: J. Appl. Spectrosc. 34, 627 (1981)]

12 R.I. Personov: in Spectroscopy and Excitation Dynamics of Condensed Molecular Systems, ed. by V.M. Agranovich, R.M. Hochstrasser (North-Holland, Amsterdam 1983) p. 555; G.J. Small, ibid., p. 515

13 K.K. Rebane: Zh. Prikl. Spektrosk. 37, 906 (1982) [English transl.: J. Appl. Spectrosc. 37, 1346 (1982)]; K.K. Rebane: J. Lumin. 31/32, 744 (1984); Cryst. Latt. Def. and Amorph. Mat. 12, 427 (1985); J. Friedrich, D. Haarer: Angew. Chem. 23, 113 (1984)

14 W.J. Weber (ed.): Optical Linewidths in Glasses, J. Lumin. 36, N 4/5 (1987)

15 A.K. Rebane, R.K. Kaarli, P.M. Saari: Pis'ma Zh. Eksp. Teor. Fiz. 38, 320 (1983) [English transl.: JETP Lett. 38, 383 (1983)]; Opt. Spektrosk. 55, 405 (1983) [English transl.: Opt. Spectrosc. USSR 55, 238 (1983)]; P. Saari, A. Rebane: Eesti NSV Tead. Akad. Toim. Füüs. Matem (Proc. Estonian SSR Acad. Sci.) 33, 322

(1984) (in Russian); A. Rebane, R. Kaarli: Chem. Phys. Lett. 101, 317 (1983); P.M. Saari, R.K. Kaarli, A.K. Rebane: Kvantovaya Elektron. 12, 672 (1985) [English transl.: Sov. J. Quantum Electron. 15, 443 (1985)]
16 E.V. Shpol'skii: Usp. Fiz. Nauk 71, 215 (1960); 77, 32 (1962) [English transl.: Sov. Phys. - Usp. 5, 522 (1962)]; 80, 755 (1963) [English transl.: Sov. Phys. - Usp. 6, 255 (1963)]
17 E.F. Gross, B.S. Razbirin, S.A. Permogorov: Dokl. Akad. Nauk SSSR 147, 338 (1962); E.F. Gross, S.A. Permogorov, B.S. Razbirin: 154, 1306 (1964)
18 E.D. Trifonov: Dokl. Akad. Nauk SSSR 147, 826 (1962)
19 M.A. Krivoglaz, S.I. Pekar: in Trudy Instituta Fiziki AN Ukr. SSR, Vol. 4 (Kiev 1953) p. 37
20 R. Dicke: Phys. Rev. 89, 472 (1953); P. Wittke, R.H. Dicke: Phys. Rev. 103, 620 (1956)
21 P.P. Feofilov: in Materialy V soveshchania po luminetstsentsii, Tartu, 1956, ed. by Ch. B. Lushchik (Estonian SSR Acad. Sci., Tartu 1957) p. 3; A.A. Kaplyanskii: Opt. Spektrosk. 6, 424 (1959); 7, 677 (1959); 10, 165 (1961) [English transl.: Opt. Spectrosc. USSR 10, 83 (1961)]; A.A. Kaplyanskii, P.P. Feofilov: Opt. Spektrosk. 16, 264 (1964) [English transl.: Opt. Spectrosc. USSR 16, 144 (1964)]
22 A.L. Schawlow: in Advances in Quantum Electronics, ed. by J.R. Singer (New York 1961) p. 50
23 K.K. Rebane, V.V. Hizhnyakov: Opt. Spektrosk. 14, 362 (1963); 14, 491 (1963) [English transl.: Opt. Spectrosc. USSR 14, 193 (1963); 14, 262 (1963)]
24 K.K. Rebane: Opt. Spektrosk. 16, 594 (1964) [English transl.: Opt. Spectrosc.USSR 16, 324 (1964)]
25 K. Rebane, P. Saari, T. Tamm: Eesti NSV Tead. Akad. Toim. Füüs. Matem. (Proc. Estonian SSR Acad. Sci.) 19, 251 (1970) (in Russian)
26 K. Rebane, L. Rebane: Eesti NSV Tead. Akad. Toim. Füüs. Matem. Tehn. (Proc. Estonian SSR Acad. Sci.) 14, 309 (1965) (in Russian)
27 K.K. Rebane, L.A. Rebane: Pure and Appl. Chem. 37, 161 (1974); A.M. Freiberg, L.A. Rebane: in this issue
28 R.A. Avarmaa, K.K. Rebane: Spectrochim. Acta A41, 1365 (1985); R.A. Avarmaa: in this issue; R.A. Avarmaa, K.K. Rebane: Usp. Fiz. Nauk (1987) in press
29 R. Avarmaa, I. Renge, K. Mauring: FEBS Lett. 167, 186 (1984); I. Renge, K. Mauring, R. Avarmaa: Biochim. Biophys. Acta 766, 501 (1984)
30 K.K. Rebane, A.A. Gorokhovskii: J. Lumin. 36, 237 (1987); A.A. Gorokhovskii: in this issue
31 V. Hizhnyakov, I. Rebane: Eesti NSV Tead. Akad. Toim. Füüs. Matem (Proc. Estonian SSR Acad. Sci.) 26, 260 (1977) (in Russian); V. Hizhnyakov: Tech. Report ISSP. University of Tokyo, Ser. A. N 860 (1977); I.K. Rebane, A.L. Tuul, V.V. Hizhnyakov: Zh. Eksp. Teor. Fiz. 77, 1302 (1979); V.V. Hizhnyakov, I.K. Rebane: Eesti NSV Tead. Akad. Toim. Füüs. Matem (Proc. Estonian SSR Acad. Sci.) 35, 406 (1986) (in English)
32 V.V. Hiznnyakov: in this issue
33 K.K. Rebane, R.A. Avarmaa, A.A. Gorokhovskii: Izv. Akad. Nauk SSSR, Ser. Fiz. 39, 1794 (1975) [English transl.: Bull. Acad. Sci. USSR, Phys. Ser. 39, 8 (1975)]; E.I. Al'shits, R.I. Personov, V.I. Stogov: Izv. Akad. Nauk SSSR, Ser. Fiz. 39, 1918 (1975) [English transl.: Bull. Acad. Sci. USSR, Phys. Ser. 39, 119 (1975)]
34 T.B. Tamm, J.V. Kikas, A.E. Sirk: Zh. Prikl. Spektrosk. 24, 315 (1976)
35 K.K. Rebane, L.A. Rebane: in Persistent Spectral Hole Burning. Science and Applications, ed. by W.E. Moerner (Springer) in press
36 E. Realo: in this issue and references therein
37 A. Winnacker, R.M. Shelby, R.M. Macfarlane: Opt. Lett. 10, 350 (1985)
38 K.K. Rebane: Eesti NSV Tead. Akad. Toim. Füüs. Matem. (Proc. Estonian SSR Acad. Sci.) 34, 441 (1985) (in Russian)
39 K.K. Rebane: in Molecular Electronics, Proc. 4th Int. School on Condensed Matter Physics, Varna, Sept. 18-27, 1986, ed. by M. Borissov (World Scientific, Singapore 1987) p. 431

Homogeneous Broadening of Zero-Phonon Lines in the Impurity Spectra of Crystals and Glasses

M.A. Krivoglaz

Institute of Metal Physics, Ukrainian SSR Acad. Sci.
SU-252642 Kiev 142, Ukrainian SSR, USSR

Abstract

The broadening of zero-phonon lines (ZPL) in the light absorption spectra of impurity crystals, caused by the interaction of the electrons of impurity centres with phonons and localized vibrations, is considered. The peculiarities of the homogeneous broadening of ZPL in the impurity spectra of glasses, studied by selective spectroscopy methods, are discussed. The resulting average spectral distribution of a ZPL that is broadened by the interaction with low-frequency two-level modes or quasilocal vibrations has been found. It is shown that in glasses the ZPL broadening, Ω, is usually caused by dynamical shifts of electronic levels in fluctuational transitions between the modes' states during the lifetime of the excited electronic state or during the hole existence. The temperature dependence of Ω is connected with the decrease of the concentration of excited modes with the fall of temperature. The dependences $\Omega(T)$ obtained allow one to explain the experimental results on ZPL in glasses.

1. Zero-Phonon Lines in the Spectra of Impurity Centres in Perfect Crystals

The interaction of the electrons of impurity centres with the vibrations of the atoms of solids has an essential effect on impurity emission and absorption spectra and leads to the formation of complex vibronic spectra. A quantitative theory of these spectra was first developed in the investigations by *Pekar* [1,2] as well as by *Huang Kun and Rhys* [3] within the frames of adiabatic approximation. In this approximation, the spectral distributions of optical vibronic transitions between non-degenerate electronic levels are determined by the difference of the adiabatic Hamiltonians $H_{s'}$ and H_s of the final and initial electronic states s' and s:

$$H_{s'} - H_s - \omega_{s's} = \sum_{\kappa} V_{\kappa}^{(1)} a_{\kappa} + \sum_{\kappa\kappa'} (V'_{\kappa\kappa'} a_{\kappa} a_{\kappa'} + V_{\kappa\kappa'} a_{\kappa}^{+} a_{\kappa'}) + \text{h.c.} \quad . \tag{1}$$

Here $\hbar = 1$, a_{κ}^{+}, a_{κ} are the operators of the creation and annihilation of the quanta of normal vibrations in a crystal containing an impurity centre in the initial state s, $\omega_{s's} = \omega_{s'} - \omega_s$ is the difference of constant terms of $H_{s'}$ and H_s.

In the first papers [1-3], the simplest case of an electron-phonon interaction, linear in a_κ, a_κ^+, in the absence of the dispersion of vibration frequencies ω_κ (when all $\omega_\kappa = \omega_0$), was considered. In this limiting case, the spectrum consists of equidistant lines separated by ω_0, corresponding to phototransitions with various numbers of absorbed and emitted phonons. The envelope of the resulting linear spectrum is of Gaussian shape at high temperatures and at low temperatures it has a characteristic shape called "a Pekarian".

A more general theory of impurity spectra considering the dispersion of vibration frequencies enabled the investigation of the shape of spectral distributions as well as their fine structure [4]. In this work, it was shown that even in the case of an essential electron-phonon interaction, along with comparatively broad bands of electron-phonon transitions, extremely narrow zero-phonon lines (ZPL) of purely electronic transitions must exist in the spectra. The widths of the latter are much smaller than those of spectral lines in gases and they appear to be optical analogs of Mössbauer lines.

In the case of a harmonic crystal and a linear electron-phonon interaction [4], the latter, in the case of the absence of radiationless transitions, does not broaden ZPL(they have only natural width) but only weakens their intensity by some factor

$$\exp(-W) = \exp[-\sum_\kappa |V_\kappa^{(1)}|^2 \omega_\kappa^{-1}(1 + 2n_\kappa)] \quad , \qquad (2)$$
$$n_\kappa = [\exp(\omega_\kappa/\kappa T) - 1]^{-1} \quad .$$

This is an analog of the Debye-Waller factor and it is of the same structure. It depends, however, not only on the characteristics of the phonon spectrum but also on the constants of electron-phonon interaction.

The quadratic electron-phonon interaction (the terms $V_{\kappa\kappa'} a_\kappa^+ a_{\kappa'}$ in (1)) as well as the anharmonicity of vibrations modulate the frequency of the electronic transition and lead to the broadening of ZPL. To determine the broadening it is convenient to proceed from a general integral expression for the cross-section $\sigma(\omega - \omega_{s's})$ of the absorption of light of the frequency ω by the impurity centre [5]. In the adiabatic approximation, in the case of transition between singlet levels

$$\sigma(\omega - \omega_{s's}) = C \int_{-\infty}^{\infty} dt \exp[i(\omega - \omega_{s's})t - g_0(t) - \gamma_0|t|/2] \quad ,$$

$$\exp(-g_0(t)) = \langle \exp(iH_s t) \exp(-i(H_s + \Delta H)t) \rangle = \langle T \exp[-i\int_0^t \Delta H(t_1)dt_1] \rangle \quad , \qquad (3)$$

$$\Delta H = H_{s'} - H_s - \omega_{s's} \quad .$$

Here C practically does not depend on ω within the limits of a ZPL, γ_0 is the natural width, $\langle\ldots\rangle$ denotes the averaging over the states of the medium ($\beta = 1/\kappa T$) with the weight factor $\exp(-\beta H_s)$, T is the symbol of chronological ordering.

In case of sufficiently weak interaction with the vibrations of the continuous spectrum it is easy to calculate the function $g_0(t) = g_{ph}(t)$ for a harmonic crystal and, consequently, the broadening and shift of the ZPL with the accuracy up to quadratic terms. The part $g'_{ph}(t)$ of $g_{ph}(t)$, determining the ZPL broadening (the other terms in $g_{ph}(t)$ determine the maximum position and intensity of the ZPL as well as its phonon wings), is of the following form [6]:

$$g'_{ph}(t) = \sum_{\kappa\kappa'} |V_{\kappa\kappa'}|^2(\omega_\kappa - \omega_{\kappa'})^{-2}(n_\kappa + 1)n_\kappa[1 - \exp(-i(\omega_\kappa - \omega_{\kappa'})t)] \quad . \tag{4}$$

In case of large times $|t| >> \overline{\omega_\kappa^{-1}}$

$$g'_{ph}(t) = \gamma'|t|/2 + i\tilde{V}'t \quad , \qquad \tilde{V}' = \sum_\kappa V_{\kappa\kappa}n_\kappa + O(V^2) \quad ,$$
$$\gamma' = \pi \sum_{\kappa\kappa'} |V_{\kappa\kappa'}|^2(n_\kappa + 1)n_\kappa\delta(\omega_\kappa - \omega_{\kappa'}) \quad , \tag{5}$$

i.e. interaction with phonons leads to the broadening of ZPL. In this case the ZPL has a Lorentzian shape and its width at the half-height is $\gamma_0 + \gamma'$ [7-13]. The broadening γ' is proportional to T^2 at high temperatures and diminishes rapidly according to T^7 at $T \to 0$. Expression (5) for γ' becomes inapplicable as soon as the quadratic electron-phonon interaction is not weak. Calculations generalizing the theory to this case have been carried out [14,15].

If the anharmonicity of vibrations is allowed for, the broadening of ZPL has already been caused by electron-phonon interaction linear in phonon operators [9-13]. The corresponding contribution to γ' is determined by a formula similar to (5), only $V_{\kappa\kappa'}$ must be replaced by a combination of $V_\kappa^{(1)}$ and anharmonicity constants.

If in the vicinity of an impurity centre there exist local or quasi-local vibrations, then their contribution into ZPL broadening is essential or sometimes even decisive [9,10]. The corresponding contribution to the integral width of ZPL in the absorption spectrum, $\Omega/2$, depends on the relation of Ω and the damping of the localized modes Γ_κ. It is described by the formula

$$\Omega = \pi \sum_\kappa V_\kappa^2\Gamma_\kappa^{-1}(n_\kappa + 1)n_\kappa \qquad (\Omega << \Gamma_\kappa) \quad ,$$
$$\Omega^2 = \xi^2 \sum_\kappa V_\kappa^2 (n_\kappa + 1)n_\kappa \qquad (\Omega >> \Gamma_\kappa) \quad . \tag{6}$$

Here the summing over κ is performed only for localized modes. For these modes $V_\kappa \equiv V_{\kappa\kappa}$ (see (1)) and $\xi \sim 1$. In case $\Omega >> \Gamma_\kappa$ localized modes do not manage to change their quantum states during the characteristic time Ω^{-1} and the ZPL is the superposition of the lines which are displaced by the distance V_κ and which correspond to various quantum numbers of modes. This superposition may have a smeared fine structure which disappears at $\Gamma_\kappa \sim V_\kappa$. If, however, $\Omega << \Gamma_\kappa$, then during Ω^{-1} transitions between the states of modes can take place and a dynamical narrowing of the modulation spectrum takes place. At $\Omega << \Gamma_\kappa$ the ZPL has a Lorentzian shape, while at

$\Omega \gg \Gamma_\kappa$, generally speaking, a complex shape[1]. If, however, a considerable contribution to Ω is made by a large number of localized modes, $\nu \gg 1$, then at $\Omega \gg \Gamma_\kappa$ the smoothed ZPL acquires a Gaussian shape and the constant ξ equals $2\sqrt{2\pi}$.

Broadening (6) is essential in the temperature region $\kappa T > \omega_\kappa/2$ and it diminishes exponentially at $T \to 0$. As the damping Γ_κ is usually small ($\Gamma_\kappa \ll \omega_\kappa$), the contribution of localized modes to Ω can be quite large, especially in the case of low-frequency quasi-local vibrations, and it may considerably exceed the phonon contribution (5).

Along with the modulational broadening of ZPL according to (5) and (6), the interaction with vibrations leads also to a broadening conditioned by radiationless transitions, which is especially significant in the case of small Bohr frequencies or multiplet spectra. This broadening is related to the finite lifetime of electronic states, it is determined by the probability of radiationless transitions and will further be considered as included into γ_0.

2. Physical Nature of Homogeneous and Inhomogeneous ZPL Broadenings in Disordered Media

The homogeneous broadening of ZPL considered above is conditioned by a dynamical modulation of the frequencies of electronic transitions and by radiationless transitions. Besides, in solids always a considerable inhomogeneous broadening takes place which is connected with the static scattering of transition frequencies in a real inhomogeneously-distorted sample. It often exceeds the homogeneous broadening by several orders of magnitude [12], especially in strongly-disordered media, e.g. in glasses. This circumstance tangibly complicates the obtaining of anomalously narrow ZPL lines and the studying of their homogeneous widths.

These difficulties could be overcome by special methods of selective spectroscopy. They include the methods of a selective laser excitation of the fluorescence of a selected group of impurity centres with an almost fixed transition frequency [16,17], hole burning in the absorption spectrum [18-21] and photon echo [22]. As in these experiments a group of centres with a fixed transition frequency is involved, the inhomogeneous broadening conditioned by the static shifts of frequencies appears to be excluded. As a result, the ZPL with only homogeneous broadening Ω can be distinguished and the latter investigated. For impurity centres in crystals the value and temperature dependence of Ω are in agreement with the theoretical results discussed above.

Recently, much attention has been paid to the investigation of the homogeneous broadening of ZPL in amorphous matter. At high temperatures it is qualitatively the same as in crystals, at low temperatures, however, Ω exceeds the corresponding value

1 The shape and the fine structure of the spectral distribution of the ZPL in a general case of an arbitrary V_κ/Γ_κ have been considered in [42].

in crystals by several orders of magnitude and disappears much slowlier. Thus, for the impurity centres Eu^{3+} in silicon glasses the law $\Omega \sim T^{1.8 \pm 0.2}$ was obtained in the interval $7\ K < T < 80\ K$ [23] and proved for $T = 1.6$ K [24]. For Pr^{3+} in amorphous substances BeF_2 and GeO_2 it was found that $\Omega \sim T^{1.85 \pm 0.2}$ in a wide temperature region from 8 to 300 K [25], while for Pr^{3+} in silicon glasses $\Omega \sim T$ at $1.6\ K < T < 20\ K$ [24]. In some organic glasses, Ω depends on T linearly [26], according to the law $T^{1.5 \pm 0.3}$ [21], in proportion to $T^{1.3 \pm 0.1}$ ($0.4\ K < T < 20\ K$) [27] and in some cases, it tends to the nonzero limit at $T \to 0$ [28]. In amorphous polymer, investigated at extremely low temperatures, $0.05\ K < T < 1.5\ K$, complicated dependences $\Omega(T)$ were obtained, which at some regions are approximated by power functions with exponents $\sim$1.0-1.5 [29].

The fact that the peculiarities of homogeneous broadening in glasses are manifested at low temperatures indicates that they are connected with low-temperature modes. As is well known, in glasses a significant role is played by two-level modes, the conception of which was introduced for explaining the thermal properties of glasses [30,31]. Immediately after the peculiarities of homogeneous broadening in glasses had been detected, the suppositions were expressed [23,32-35] that these could be explained by the interaction of impurity electrons with two-level modes. However, the investigations developing the qualitative theory [33-35] did not usually allow for the basic mechanism of modulational broadening, conditioned by the shifts of the electronic levels of centres on fluctuational transitions of two-level modes during the lifetime of the excited state in case of selective fluorescence or in the interval between the burning of a hole and its investigation. Besides, in these studies, the ZPL broadening was determined as the mean $\bar{\Omega}$ of the widths taken at various values of mode parameters. In a consistent theory, it is necessary to average over the partial spectral distributions. The width Ω of the resulting curve may notably differ from $\bar{\Omega}$ and have a different temperature dependence. In connection with this, it was necessary to calculate the broadening conditioned by the interaction of impurity centres with two-level modes by means of a more consistent averaging for various broadening mechanisms, the basic one referred to included. Such calculations have been performed independently by somewhat different methods in [36,37].

A significant contribution to the broadening may be made also by quasi-local vibrations with sufficiently small ω_κ and $\Gamma_\kappa \sim \omega_\kappa^4$ (see formula (6)). In the cases when the low-temperature broadening in glasses considerably exceeds the broadening in crystal modifications of the same matrix, the probability of its being connected with the quasi-local vibrations localized at the impurity centre itself (in case of small differences of the short-range order in glass and crystal such vibrations should not appear in most of the centres in glasses) is evidently small. However, there exist arguments (see, e.g. [38,37]) indicating that in glasses, regardless of the presence of impurity centres, there exists an appreciable density of modes of quasi-local vibrations ($\sim 10^{-2}$-10^{-3} atom densities) in the sites where force constants are weakened due to the fluctuations of interatomic distances. A long-range interaction of such modes with the electrons of centres should lead to the contribution to the ZPL broadening considered in [37].

Thus, in the model regarded below, low-temperature (two-level or quasi-local) modes are statistically distributed in glasses at various random distances up to the given impurity centre. The partial spectra, corresponding to various positions of centres or (which is the same) to smeared mode positions and parameters, should then be averaged out, giving the resulting spectrum which is investigated experimentally.

The resulting effects depend to a considerable degree on the relation of three times: the lifetime t_0 of the excited electronic state in the method of selective fluorescence of the time of hole existence in the spectrum, the mode relaxation time $t^0 \sim \Gamma_\kappa^{-1}$ and the characteristic time Ω^{-1}, which determines the interval making the main contribution to integral (3) for the spectral representation $\sigma(\omega - \omega_{s's})$. Further we shall limit ourselves to a case when the time t_0 is the largest, i.e.

$$t_0 \gg t^0 \sim \Gamma_\kappa^{-1} , \qquad t_0 \gg \Omega^{-1} \tag{7}$$

(the contribution to the homogeneous broadening of the modes, for which $t_0 \ll t^0$, has essentially decreased and at $t_0 \to 0$ they contribute only to inhomogeneous broadening). The relation of Γ_κ and Ω may be arbitrary.

If

$$\Gamma_\kappa \ll \Omega \quad , \tag{8}$$

then in the time Ω^{-1} localized modes do not manage to change their states. Owing to fluctuations, the corresponding quantum numbers are different and the related various shifts of lines in the impurity spectrum lead to the appearance of a set of lines. When overlapping, they form a broadened resulting spectrum and, on investigating the ordinary absorption spectrum, at $\Gamma_\kappa \ll \Omega$ this mechanism makes a contribution to inhomogeneous broadening. However, as $t^0 \ll t_0$, the values of quantum numbers change during the excited state lifetime or during the existence of the hole, so that for the given impurity centre the shifts in the spectra of excitation and emission (or repeated absorption) acquire various, randomly differing values. When using the methods of selective spectroscopy this contribution to broadening is not excluded and is interpreted as homogeneous broadening (that at $\Gamma_\kappa \ll \Omega$ is not related to the finite lifetimes of quantum states and that is of a modulational nature).

In the opposite case,

$$\Gamma_\kappa \gg \Omega \tag{9}$$

during the time Ω^{-1} the state of modes alters repeatedly and substantial dynamical decrease of the considered modulational ZPL broadening takes place (cf. with formula (6)). In glasses, there occurs a large distribution of mode parameters and, as a rule, even if condition (9) is fulfilled for the majority of them, for some group of modes Γ_κ is small and condition (8) is fulfilled. Essential is that, as shown in [37], it is the last group of modes that usually makes the main contribution to the homogeneous ZPL broadening, i.e. even in this case the main role is played by the above-mentioned modulational mechanism of broadening by modes with $\Gamma_\kappa \lesssim \Omega$ (if such modes constitute an appreciable part). Analogously, if for some group of modes $t^0 \sim \Gamma_\kappa^{-1} \gg t_0$, then their contribution to Ω is small and the main contribution to homogeneous broadening comes from modes with $t^0 < t_0$. Note that this broadening mechanism is analogous to that of spectral diffusion [39], the theory of which was applied to the investigation of the mode-conditioned ZPL broadening in [36]. The influence of the mechanism under consideration on the spectrum of resonant light scattering, in particular, in the case of $t^0 \gg t_0 \gg \Omega^{-1}$, was semiphenomenologically discussed in [40].

3. Spectral Distributions of ZPL in Glasses, Determined by the Method of Selective Spectroscopy

To study the influence of localized modes on ZPL it is first necessary to find the cross-section of the absorption of light by a single impurity centre in case of fixed mode positions and then perform averaging over mode positions and characteristics. It is convenient to subdivide the medium into small elements of the volume v_0 (of the order of an atomic volume) and the possible values of the parameters κ, characterizing the modes (e.g. their frequencies ω_κ), into small intervals. The statistical properties of modes can be uniquely characterized by the random quantities $c_{\underline{r}\kappa}$ which acquire values 1, 0 if in the $\underline{r}$-th volume element the mode with the parameters falling into the given interval κ is either present or absent, respectively. Considering that the concentration of modes, c, is small, we shall further neglect their mutual interaction as well as their interaction with the impurity centre, i.e. we shall regard the numbers $c_{\underline{r}\kappa}$ as independent random quantities.

The adiabatic Hamiltonian of the impurity centre H_s depends on the positions and characteristics of the neighbouring modes. In case of small c and weak interaction of the electrons of the centre with the medium, H_i, it can be written as follows:

$$H_s = H_0 + H_i^s \,, \qquad H_i^s = H_{iph}^s + \sum_{\underline{r}\kappa} c_{\underline{r}\kappa} H_{\underline{r}\kappa}^s \,. \tag{10}$$

Here summing has been performed over all the volume elements $\underline{r}$ and the intervals κ, H_0 are the terms of the adiabatic Hamiltonian of the centre and the medium, which are independent of the electron state, s, $H_{iph}^s = (s|H_{iph}|s)$, $H_{\underline{r}\kappa}^s = (s|H_{\underline{r}\kappa}|s)$, H_{iph} and $H_{\underline{r}\kappa}$ are the Hamiltonians of the interaction of the electrons of the centre with phonons and with the modes in the positions $\underline{r}$ with the parameters κ.

Substituting expression (10) into (3) and performing the averaging <...> over the initial states of the medium at fixed $c_{\underline{r}\kappa}$ with the weight-factor $\exp(-\beta H_s)$, the cross-section of absorption, σ, by a single impurity centre can be found. In case of small c the interaction between modes and their influence on phonons can be neglected. In this case, $\exp(-g_0(t))$ in (3) is decomposed into the product of independent factors with $\underline{r}\kappa$ and, as $c_{\underline{r}\kappa} = 1, 0$,

$$g_0(t) = g_{ph}(t) + \sum_{\underline{r}\kappa} c_{\underline{r}\kappa}\, g_{\underline{r}\kappa}(t) \quad . \tag{11}$$

Here $g_{ph}(t)$ and $g_{\underline{r}\kappa}(t)$ are obtained from formula (3) for $g_0(t)$ by replacing H_s with $H_0 + H^S_{iph}$ or with $H_0 + H^S_{\underline{r}\kappa}$, respectively.

The spectrum of the narrow ZPL is determined by the behaviour of $g_0(t)$ at large $t \sim \Omega^{-1}$. For this region it is easy to obtain explicit expressions for $g_{\underline{r}\kappa}(t)$ within the limiting cases (8) and (9).

If $\Gamma_\kappa \ll \Omega$, then during $t \sim \Omega^{-1}$ the localized modes cannot change their states corresponding to the Hamiltonian $H_0 + H^S_i$. These can be characterized by the quantum numbers $n_{\underline{r}\kappa}$, where $n_{\underline{r}\kappa} = 0, 1$ for two-level modes and $n_{\underline{r}\kappa} = 0, 1, 2, \ldots$ for quasi-local vibrations. In case the interaction of $H_{\underline{r}\kappa}$ with the electrons of the centre is weak, the energy change of these states can be calculated in the first approximation of the perturbation theory (on transitions corresponding to ZPL, $n_{\underline{r}\kappa}$ do not change) and characterized by the constants $V_{\underline{r}\kappa}$ depending on the distance between the mode and the centre:

$$\begin{aligned} &(n_{\underline{r}\kappa}|\Delta H_{\underline{r}\kappa}|n_{\underline{r}\kappa}) \equiv (s'n_{\underline{r}\kappa}|H_{\underline{r}\kappa}|s'n_{\underline{r}\kappa}) - (sn_{\underline{r}\kappa}|H_{\underline{r}\kappa}|sn_{\underline{r}\kappa}) = V_{\underline{r}\kappa}\,(n_{\underline{r}\kappa} + \text{const}) \quad , \\ &|V_{\underline{r}\kappa}| \ll \omega_\kappa, \kappa T \quad . \end{aligned} \tag{12}$$

Calculating $g_{\underline{r}\kappa}(t)$ according to (3) by summing over $n_{\underline{r}\kappa}$, we obtain

$$\begin{aligned} &g_{\underline{r}\kappa}(t) = -\ln h_{\underline{r}\kappa}(t) \quad , \qquad h_{\underline{r}\kappa}(t) = \sum_n w_\kappa(n)\exp(-iV_{\underline{r}\kappa}nt) \quad , \\ &\Gamma_\kappa^{-1} \gg |t| \quad , \qquad w_\kappa(n) = \exp(-\beta\omega_\kappa n)\,[\sum_n \exp(-\beta\omega_\kappa n)\,]^{-1} \quad , \\ &n_\kappa = \langle n_{\underline{r}\kappa} \rangle = \sum_n n w_\kappa(n) \quad . \end{aligned} \tag{13}$$

Here, for two-level modes in the sums $n \equiv n_{\underline{r}\kappa} = 0, 1$, for quasi-local vibrations $n = 0, 1, 2, \ldots$.

In the opposite limiting case, $\Gamma_\kappa \gg \Omega$, in integral (3), large times $|t| \gg t^0 \sim \Gamma_{\underline{r}\kappa}^{-1}$ are essential. In this region, for $g_{\underline{r}\kappa}(t)$ and $g_{ph}(t)$ (in case of phonons $t^0 \sim \omega_{\underline{\kappa}}^{-1}$ and at weak coupling always $t^0 \ll \Omega^{-1}$) simple asymptotic expressions (see, e.g. [10]) are valid and neglecting small (at small H_i) constant terms

$$\begin{aligned} &g_{\underline{r}\kappa}(t) = \gamma_{\underline{r}\kappa}|t|/2 + i\tilde{V}_{\underline{r}\kappa}t \quad , \quad |t| \gg \Gamma_\kappa^{-1} \quad , \quad \tilde{V}_{\underline{r}\kappa} = V_{\underline{r}\kappa}\, n_\kappa + O(V^2) \quad , \\ &\gamma_{\underline{r}\kappa} = \lim_{\omega\to 0} \int_{-\infty}^{\infty} \langle(\Delta H_{\underline{r}\kappa}(t) - \langle\Delta H_{\underline{r}\kappa}\rangle)\,(\Delta H_{\underline{r}\kappa}(0) - \langle\Delta H_{\underline{r}\kappa}\rangle)\rangle \exp(-i\omega t)\, dt \quad . \end{aligned} \tag{14}$$

The phonon contribution $g_{ph}(t)$ (see (5)) to $g_0(t)$ is determined by formula (14), even if for $g_{\underline{r}\kappa}(t)$ expression (13) is valid. In the same way, it is possible to show that the ZPL of the $s' \to s$ transition in the emission spectrum is described by formula (3) with the same function, $g_0(t)$, i.e. its intensity is proportional to $\sigma(\omega - \omega_{s's})$.

Let us consider the spectral distribution $I(\omega - \omega_0, \omega_0)$ of ZPL in the spectrum of a selective fluorescence excited by a laser with the fixed frequency ω_0. The scattering of the frequencies $\delta\omega_{s's}$ of the electronic transitions $\omega_{s's}$ of various centres, i.e. inhomogeneous broadening, usually exceeds the homogeneous broadening Ω by several orders of magnitude. Therefore, light absorption with the frequency ω_0 results in the excitation of centres whose frequencies $\omega_{s's} \approx \omega_0$ have a distribution function which is proportional to $\sigma(\omega_0 - \omega_{s's})$ with the proportionality coefficient weakly depending on ω_0. On light emission, each excited centre makes a contribution to the ZPL of the fluorescence spectrum, which is proportional to $\sigma(\omega - \omega_{s's})$, i.e. the contribution of a given centre with the Bohr frequency, $\omega_{s's}$, to the resulting distribution $I(\omega - \omega_0, \omega_0)$ is proportional to $\sigma(\omega_0 - \omega_{s's})\sigma(\omega - \omega_{s's})$. Here, every multiplier is determined by formula (3). In the case under consideration, when condition (7) is fulfilled, the lifetime of the excited electronic state t_0 considerably exceeds the relaxation time of modes, t^0, so that their thermal distributions on light absorption and emission are not correlated and the averagings <...> over the mode states in these multipliers are carried out independently.

Therefore, after averaging over the centres [21,37]

$$I(\omega - \omega_0, \omega_0) = \text{const} \langle \int \sigma(\omega_0 - \omega_{s's})\, \sigma(\omega - \omega_{s's})\, d\omega_{s's} \rangle_c$$

$$= (I_i/2\pi) \int_{-\infty}^{\infty} dt \exp(i\Delta\omega - g(t)) \quad , \qquad \Delta\omega = \omega - \omega_0 \quad , \tag{15}$$

$$\exp(-g(t)) = \langle \exp(-g_0(t) - g_0(-t) - \gamma_0|t|) \rangle_c \quad .$$

Here I_i is the integral ZPL intensity, $\langle\ldots\rangle_c$ denotes averaging over the centres or, which is equivalent to it, over the configurations and parameters of the modes around the centre, i.e. over statistically independent variables $c_{\underline{r}\kappa}$. Considering that $c_{\underline{r}\kappa}$ acquire the values $c_{\underline{r}\kappa} = 1$, with the probability cp_κ, and $c_{\underline{r}\kappa} = 0$, with the probability $1 - cp_\kappa$, where $c = v_0 N_0 \ll 1$ is the mode concentration (N_0 is their number in unit volume) and p_κ, the density of the probability that their parameters acquire the values κ, and averaging over $c_{\underline{r}\kappa}$ with consideration for (11), we find that

$$g(t) = \gamma' |t| + c \sum_{\underline{r}\kappa} p_\kappa \{1 - \exp(-g_{\underline{r}\kappa}(t) - g_{\underline{r}\kappa}(-t))\} \quad ,$$

$$\sum_\kappa p_\kappa = 1, \qquad c \ll 1 \quad . \tag{16}$$

Here the natural width γ_0 is included into γ'.

An analogous reasoning shows that formulae (15) and (16) also describe ZPL in a spectrum obtained by the hole burning method in case of weak excitation and small thickness of the sample (a more general case has been discussed in review article [21], and in the papers referred to therein).

In case of large mode relaxation time $t^0 \sim \Gamma_\kappa^{-1} >> \Omega^{-1}$ (but $t^0 << t_0$) it follows from formulae (16) and (13) that

$$g(t) - \gamma'|t| = c \sum_{\underline{r}\kappa} p_\kappa [1 - h_{\underline{r}\kappa}(t)\, h_{\underline{r}\kappa}(-t)]$$
$$= c \sum_\kappa p_\kappa \sum_{n,n'} w_\kappa(n)\, w_\kappa(n') \sum_{\underline{r}} [1 - \cos V_{\underline{r}\kappa}(n - n')t] \quad , \qquad (17)$$
$$\Omega >> \Gamma_\kappa >> t_0^{-1} \quad .$$

Usually $\Delta H_{\underline{r}\kappa}$, $V_{\underline{r}\kappa}$ and $\gamma_{\underline{r}\kappa}$ depend by power law on the distance between the centre and the localized mode as follows:

$$\Delta H_{\underline{r}\kappa} \sim 1/r^\kappa \quad , \qquad V_{\underline{r}\kappa} = V_\kappa (r_0^\kappa / r^\kappa) \quad ,$$
$$\gamma_{\underline{r}\kappa} = \gamma_\kappa (r_0^{2\kappa}/r^{2\kappa}), \qquad r_0 = v_0^{1/3} \quad . \qquad (18)$$

In this case, it is easy to sum over $\underline{r}$ and over n, n' in (17) to find a simple expression for g(t):

$$g(t) = \gamma'|t| + [\frac{\Gamma(1 + \alpha^{-1})}{\pi} \Omega|t|]^\alpha \quad , \qquad \Omega >> \Gamma_\kappa >> t_0^{-1} \quad ,$$
$$\Omega^\alpha = \frac{2\pi^{2+\alpha} c}{3\Gamma(\alpha)\Gamma^\alpha(1 + \alpha^{-1}) \sin(\pi\alpha/2)} \sum_\kappa p_\kappa V_\kappa^\alpha W_\kappa(T,\alpha) \quad , \qquad (19)$$

where

$$\alpha = \frac{3}{\kappa} \le 1 \quad , \qquad W_\kappa(T,\alpha) = \sum_{n,n'} w_\kappa(n)\, w_\kappa(n')\, |n - n'|^\alpha \quad . \qquad (20)$$

For two-level modes and quasi-local vibrations

$$W_\kappa(T,\alpha) = \frac{1}{2}(\mathrm{ch}\, \frac{\beta\omega_\kappa}{2})^{-2} \quad , \qquad W_\kappa(T,\alpha{=}1) = (\mathrm{sh}\, \beta\omega_\kappa)^{-1} \quad , \qquad (21)$$

respectively.

Formulae (15) and (19) determine the distribution $I(\Delta\omega, \omega_0)$. In case of $\pi\gamma' << \Omega$ it is of the form [36,37]:

$$I(\Delta\omega, \omega_0) = (I_i/\Omega)\, f(\frac{\Delta\omega}{\Omega}, \alpha) \quad ,$$
$$f(x,\alpha) = \Gamma^{-1}(1 + \alpha^{-1}) \int_0^\infty \cos \frac{\pi x z}{\Gamma(1 + \alpha^{-1})} \exp(-z^\alpha) dz \quad . \qquad (22)$$

The integral width of this distribution equals Ω and is determined by formula (19). If the interaction constant $V_{\underline{r}\kappa} \sim r^{-3}$, then $\alpha = 1$ and distribution (22) is of a Lorentzian shape. At $\pi\gamma' \gtrsim \Omega$ its integral width equals $\pi\gamma' + \Omega$ (at $\pi\gamma' > \Omega$ this sum must be substituted for Ω in the above-presented criteria). If, however, $\kappa > 3$, then distribution (22) differs from the Lorentz' one: it is narrower in the central part and falls slower at wings.

In case of small relaxation times, $\Gamma_\kappa^{-1} \ll \Omega^{-1} \ll t_0$, according to (14) $g_{\underline{r}\kappa}(t) + g_{\underline{r}\kappa}(-t) = \gamma_{\underline{r}\kappa}|t|$ and with consideration for (15), (16), (18) [37]

$$I(\Delta\omega,\omega_0) = (I_i/\Omega)\, f(\Delta\omega/\Omega,\ \alpha/2) \quad , \qquad \gamma', t_0^{-1} \ll \Omega \ll \Gamma_\kappa \quad ,$$
$$\Omega^{\alpha/2} = \frac{4\pi\ \pi^{\alpha/2}\ \Gamma(1-\alpha/2)}{3\Gamma^{\alpha/2}(1+2\alpha^{-1})}\, c\sum_\kappa p_\kappa \gamma_\kappa{}^{\alpha/2} \ . \tag{23}$$

4. Homogeneous ZPL Broadening in Glasses

Let us first regard the homogeneous broadening Ω, which is caused by the interaction with quasi-local vibrations. To determine it according to formula (19) it is necessary to know the probability density of their parameters (particularly, of the frequencies ω_κ), p_κ, and the law of the interaction of the electrons of the centres with localized modes, i.e. κ and V_κ. The distribution p_κ can be found by using a simple model of a nonlinear oscillator for low-frequency modes in glasses [38]. In this model, such modes are described by the effective Hamiltonian $H_{0\kappa} \equiv H(x)$:

$$H(x) = -\frac{1}{2M}\frac{d^2}{dx^2} + V(x) \quad , \qquad V(x) = \frac{1}{2}V_2x^2 + \frac{1}{3}V_3x^3 + \frac{1}{4}V_4x^4 \ . \tag{24}$$

Here M is the effective reduced mass of a weakly bound atom (WBA) or of a molecule with the coordinate x; V_2 and V_3 are random quantities. To determine the probability densities $P(V_2,V_3)$ at small $|V_2|$, $|V_3|$ we take that in a one-dimensional model V is equal to the sum of the energies $\phi(x - x_i)$ of the interaction of WBA with the other atoms, i, while the probability density of the coordinates x_i, $P(x_i)$, has no peculiarities. From the condition $V'(x_0) \equiv dV/dx|_{x=x_0} = 0$ in the point of minimum, $x = x_0$, it follows that at small $|V_2|$ $\partial x_0/\partial x_i = \phi_i''/V_2$ ($\phi_i'' \equiv \partial^2\phi(x_0 - x_i)/\partial x_0^2$) may be large. Owing to such strong dependence of x_0 on x_i, the values $V_2 = V''(x_0)$, $2V_3 = V'''(x_0)$ may change significantly on the variations of x_i: $\partial V_2/\partial x_i = 2V_3V_2^{-1}\phi_i'' - \phi_i'''$, $\partial V_3/\partial x_i = 3V_4V_2^{-1}\phi_i'' - \phi_i^{IV}/2$. Therefore, on calculating

$$P(V_2,V_3) = \int P(x_i)\delta[V_2 - V_2(x_0(x_i),x_i)]\ \delta[V_3 - V_3(x_0(x_i),x_i)]\prod_i dx_i$$

two cases are possible, which correspond to various structure of glass. In the case a), all $|\phi_i''| \lesssim |V_2|$ (WBA in a large cavity) and $|\nabla V_2|$, $|\nabla V_3|$ are finite at $V_2 \to 0$, i.e. as was assumed in [38,37], $P(V_2,V_3) \to$ const at $|V_2|$, $|V_3| \to 0$. In the case b), ϕ_i'' are sign-alternating and for some i $|V_2| = |\sum_i \phi_i''| \ll \phi_i''$ at $V_2 \to 0$. Hereby $|\nabla V_3| \sim |V_2|^{-1}$ on the surface $V_3(x_0,x_i) = V_3$, but on $V_2(x_0,x_i) = V_2$ the vector ∇V_2 is almost parallel to $\nabla V_3 \perp V_3$ and its projection onto the subsurface $V_2(x_0,x_i) = V_2$,

$V_3(x_0,x_i) = V_3$ at $V_2 \to 0$, $V_3 \to 0$ is finite. Therefore, $P(V_2,V_3) = \mathrm{const}|V_2|$. Such result is also obtained in a three-dimensional model with the energies $\phi_i(\underline{r} - \underline{r}_i)$.

Analogously to [37] (where only the case a) was considered), it can be shown that the density of the probabilities of the local vibrations $\omega_\kappa = (V_2/M)^{1/2}$ is described by the formula

$$p_\kappa \equiv p_\kappa(\omega_\kappa) = D_s\omega_\kappa^s \quad , \qquad \omega_\kappa^3 >> \omega_m^3 >> \omega_4^3(1 + n_\kappa) \quad , \qquad \omega_4^3 = V_4M^{-2} \quad . \tag{25}$$

Here s = 2 in the case of a) and s = 4 in the case of b); ω_D is the Debye frequency; $D_s = (s + 1)\omega_m^{-(s+1)}$ (ω_m is the maximum frequency of quasi-local vibrations); $\omega_4 \sim 0.1\omega_D$ is the characteristic frequency at which anharmonicity becomes essential. The modes under consideration contribute to the thermal capacity as $\delta C \sim cT^{s+1}$.

As a rule, in the interaction of quasi-local vibrations with impurity centres the main role is played by an elastic interaction. It is connected with the modulation of the frequencies $\omega_{s's}$ by the low-frequency deformations u_{ij} created by these vibrations and proportional to their quantum numbers $n_{r\kappa}$. The constant of the corresponding interaction (see (12)) $V_{r\kappa} \sim r^{-3}$, i.e. in (18) $\kappa = 3$, while the coefficient V_κ has the order of magnitude (for details, see [37])

$$V_\kappa \sim \frac{\hbar V_{oom}}{10Mr_0^2\omega_\kappa G_0}\left|\frac{\partial\omega_{s's}}{\partial u_{ii}}\right| \quad . \tag{26}$$

Here V_{oom} is the coefficient in the anharmonic interaction $V_{oom}u_{ox}^2u_{mx}$ of a WBA (molecule) "0" with strongly bound neighbouring atoms m ($\underline{u}_m$ are the displacements of atoms), G_0 is the shear modulus.

The damping of quasi-local vibrations $\Gamma_\kappa \sim \omega_\kappa^4/\omega_D^3$ and is very small at small ω_κ. At low temperatures (but at $2T > \omega_4$) the modes with the frequencies $\omega_\kappa \lesssim 2T$ are excited. Therefore, at least in the region $2T < \omega_\kappa^0$ (and $\omega_\kappa < \omega_\kappa^0$) where ω_κ^0 is determined by the equality $\Gamma_\kappa(\omega_\kappa^0) = \Omega$, the condition $\Omega > \Gamma_\kappa$ is fulfilled, and for sufficiently broad ZPL it is fulfilled for all the modes with $\omega_\kappa < \omega_m$.

In this case the ZPL is described by formulae (19)-(22) for $\alpha = 1$ and it has a Lorentzian shape; its broadening according to (19), (25), (26) is equal to

$$\Omega = \frac{\pi^5}{3} D_s\xi_sLcT^s \quad , \qquad L = |V_\kappa|\omega_\kappa \quad , \qquad \xi_2 = 1 \quad , \qquad \xi_4 = \pi^2/2 \quad ,$$
$$\omega_4 < 2T < \omega_\kappa^0 \quad . \tag{27}$$

In case of sufficiently narrow ZPL at $2T > \omega_\kappa^0$, the modes with $\omega_\kappa > \omega_\kappa^0$ are also excited, the main contribution to Ω, however, is made by the modes with $\omega_\kappa < \omega_\kappa^0$ and

$$\Omega \sim \frac{2\pi^3}{3s} D_sLc(\omega_\kappa^0)^sT \quad , \qquad 2T > \omega_\kappa^0 > \omega_4 \quad , \qquad \Gamma(\omega_\kappa^0) = \Omega \quad . \tag{28}$$

At low temperatures the contribution to Ω by quasi-local vibrations is apparently essential only in the glasses of type a). Usually, the contribution by the electrical interaction with these vibrations to Ω is less than that made by the elastic one.

Let us consider now a homogeneous broadening caused by the interaction of centres with low-frequency two-level modes in glasses. These modes appear when the difference between the unperturbed energies in two adjacent potential wells Δ and the matrix element for tunnelling through the potential barrier W/2 are sufficiently small [30, 31]. With consideration for tunnelling, the energies of the resulting mixed states differ by $E = \sqrt{\Delta^2 + W^2}$. When using the creation and annihilation operators of the mixed states a_σ^+, a_σ ($\sigma = 0, 1$ correspond to the lower and upper levels) and of the phonons $a_{\underline{\kappa}}^+$, $a_{\underline{\kappa}}$ (considered in the Debye approximation), the Hamiltonian $H_{0\kappa} + H_{ph}$ of the mode $\underline{r}\kappa$, interacting with phonons, and the difference of the Hamiltonians of the states s' and s, $\Delta H_{\underline{r}\kappa}$, describing the interaction of the centre with this mode, can be written in the form (see, e.g. [35]):

$$H_{0\kappa} + H_{ph} = \sum_{\sigma=0,1} E_\sigma a_\sigma^+ a_\sigma + \sum_{\underline{\kappa}} \omega_{\underline{\kappa}} a_{\underline{\kappa}}^+ a_{\underline{\kappa}} + \sum_{\underline{\kappa}} \left(\frac{\Delta}{2E} h' + \frac{W}{2E} h''\right) \left(U_{\underline{\kappa}} a_{\underline{\kappa}} + U_{\underline{\kappa}}^* a_{\underline{\kappa}}^+\right) \quad ,$$

$$h' = a_0^+ a_0 - a_1^+ a_1 \quad , \qquad\qquad h'' = a_0^+ a_1 + a_1^+ a_0 \quad , \tag{29}$$

$$U_{\underline{\kappa}} = U(\omega_{\underline{\kappa}}/2\tilde{M}w^2)^{1/2} \exp(i\underline{\kappa r}) \quad ;$$

$$\Delta H_{\underline{r}\kappa} = \frac{1}{2} V\left(\frac{\Delta}{E} h' + \frac{W}{E} h''\right) \frac{r_0^\kappa}{r^\kappa} \quad . \tag{30}$$

Here $E_0 = 0$, $E_1 = E$, w is the sound velocity, $\tilde{M}$ is the mass of the system, Vr_0^κ/r^κ is the difference of the shifts of the transition frequencies $\omega_{s's}$ for unperturbed states of two-level modes in two potential wells, U is the difference of deformation potentials for these states (for brevity, the index κ is not written out for the quantities Δ, E, V, U, a_σ, a_σ^+, characterizing the mode κ). In the event of elastic interaction $\kappa = 3$

$$V = \chi_{ij}\, \delta p_{ij} \frac{1}{r_0^3} \quad , \qquad\qquad \chi_{ij} = \frac{\partial \omega_{s's}}{\partial u_{i'j'}} \frac{\partial^2 G_{ii'}(\underline{r})}{\partial x_j\, \partial x_{j'}} r^3 \quad ,$$

where δp_{ij} is the difference of the tensors of dipole force constants for mode states in two wells, $G_{ij}(\underline{r})$ is the Green's function of the elasticity theory, summing is performed over twice repeating indices from 1 to 3.

There exists an extensive scattering of random values Δ and $W = \omega_0 \exp(-\lambda)$ for various modes ($\exp(-\lambda)$ characterizes the overlapping of wave functions in two wells, $\omega_0 \sim \omega_D$ is the energy of zero-point vibrations [30]). The values $W_{max} > W > W_{min}$ or the values $\lambda_{min} < \lambda < \lambda_{max}$ are actual, where $W_{max} = \omega_0 \exp(-\lambda_{min}) \sim \Delta$ (at smaller λ

E increases rapidly), and λ_{max} is determined by the demand that the mode relaxation time $\tau = \Gamma^{-1} \sim \exp(2\lambda)$ (see formula (33)) be less than the characteristic observation time, in this case, t_0. An analysis of experimental data results in the estimates $\lambda_{max} \sim 10\text{-}20$, $\delta\lambda \equiv \lambda_{max} - \lambda_{min} \sim 5\text{-}10$ [30,41]. In the accepted model of two-level modes in glasses [30,31], the probability of their parameters $p_\kappa = p(\Delta,\lambda)$ has a constant value $(\Delta_m\delta\lambda)^{-1}$ for $0 < \Delta < \Delta_m \gtrsim \omega_D$ and $\lambda_{min} < \lambda < \lambda_{max}$ and equals zero beyond these intervals of Δ,λ.

The influence of two-level modes on the ZPL width, like in case of quasi-local vibrations, depends essentially on the relation between the parameters Ω, Γ and t_0^{-1}. Usually $\Gamma < 10^8\ s^{-1}$ at $T \sim 1$ K, i.e. $\Omega \gg \Gamma$ even at small $\Omega \sim 10^{-2}\ cm^{-1}$. At the same time, the condition $t_0 \gg \Gamma^{-1}$ may be fulfilled (otherwise these modes give a small contribution to the homogeneous broadening, that usually does not exceed the natural width). Therefore, the case of $\Omega \gg \Gamma \gg t_0^{-1}$, when at a weak coupling ($\Omega \ll T$) the ZPL is described by formulae (15), (19)-(22), is of main interest. According to (18), (30) in (19) $V_\kappa = V\Delta/E$. Integrating over Δ, λ with the weight factor $p_\kappa = p(\Delta,\lambda)$ in the indicated regions for Δ and λ and with function (21) for $W_\kappa(T,\alpha)$ (where $\omega_\kappa = E$), we find that the ZPL has the shape of the curve $f(\Delta\omega/\Omega,\alpha)$ (22) and the integral width [36,37]

$$\Omega = \xi V(cT/\Delta_m)^{\kappa/3} \quad , \qquad \xi = \frac{3\pi}{\kappa\Gamma(\kappa/3)}\left[\frac{2\pi^2}{3\Gamma(3/\kappa)\,\sin(3\pi/2\kappa)}\right]^{\kappa/3} \quad ,$$
$$T \gg \Omega \gg \Gamma \gg t_0^{-1} \quad . \tag{31}$$

In case of elastic or electric dipole interaction $\kappa = 3$, $\alpha = 1$, ZPL has a Lorentzian shape and, analogously to [32], $\Omega \sim cT$. For example, at $\delta p \sim r_0^3 G_0$, $c/\Delta_m \sim 10^{-2}$ eV^{-1}, $T \sim 1$ K, $|\partial\omega_{s's}/\partial u_{ii}| \sim (10^3\text{-}10^4)\ cm^{-1}$ according to (31) $\Omega \sim (10^{-3}\text{-}10^{-2})\ cm^{-1}$. For dipole-quadrupole or quadrupole-quadrupole interactions $\kappa = 4$ ($\alpha = 3/4$) or $\kappa = 5$ ($\alpha = 3/5$), the curve $f(\Delta\omega/\Omega,\alpha)$ (22) differs from the Lorentzian (narrower in the central part and falls at wings like $(\Delta\omega)^{-7/4}$ or $(\Delta\omega)^{-8/5}$) and $\Omega \sim (cT)^{4/3}$ or $\Omega \sim (cT)^{5/3}$. If $p(\Delta,\lambda)$ is not constant, but grows slowly with Δ, for example, like Δ^μ with a small μ, then the exponent $(1+\mu)\kappa/3$ in the power function $\Omega(T)$ is somewhat enhanced.

In case of sufficiently narrow ZPL for a group of modes with very large Γ the condition $\Omega \ll T$ is fulfilled. Their contribution into g(t) is determined by formulae (14), (16) and is expressed by the spectral representations of the correlators $\Delta H_{r\kappa}$, i.e. h', h", at $\omega \to 0$. The latter can be found by the Green's function method (see Appendix in [10]) and by using (14), (18), (29), (30)

$$\gamma_\kappa = \gamma'_\kappa + \gamma''_\kappa \quad , \qquad \gamma'_\kappa = 2\pi\left(\frac{V\Delta}{2E}\right)^2 \langle h',(h' - \langle h'\rangle)\rangle_{\omega\to 0}$$
$$= \frac{V^2\Delta^2}{\Gamma E^2}\exp(\beta E)\,(\exp(\beta E) + 1)^{-3} \quad ,$$

$$\gamma''_\kappa = 2\pi(\frac{VW}{2E})^2\langle h'', (h'' - \langle h''\rangle)\rangle_{\omega\to o} = \frac{8\Gamma V^2\Delta^2}{E^4}(1 + \exp(-\beta E))^{-1} \quad , \tag{32}$$

$$T \gg \Gamma \gg \Omega \gg t_0^{-1} \quad ,$$

where

$$\Gamma = \pi\sum_{\underline{\kappa}}(\frac{WU_{\underline{\kappa}}}{2E})^2 n_\kappa\delta(e - \omega_\kappa) = \frac{V^2W^2Ev_0}{16\pi Mw^5}(\exp(\beta E) - 1)^{-1} \quad . \tag{33}$$

The contribution of this group of modes to Ω has been estimated in [37] (in the same work, a comparison of formulae (32) and averaged widths with the results of [33,35] was made). The contribution of this group of modes to Ω was shown to be less than that of a group of modes with $\Gamma \ll \Omega$, which usually exists even in case of small Ω. To estimate the latter, formula (31) can be used, where under c a concentration of modes with $\Omega \gg \Gamma \gg t_0^{-1}$ is to be understood (owing to a strong dependence of $\Gamma \sim \sim \exp(2\lambda)$ on λ, the interval $\delta\lambda$ and c depend weakly on temperature).

It can be inferred from the results presented that the homogeneous broadening of impurity ZPL, Ω, in disordered media at low temperatures, investigated by the methods of selective spectroscopy, is usually mainly determined by the dynamical shifts of the frequencies of electronic transitions in the fluctuations of the quantum numbers of low-frequency localized modes during the time t_0 and is not related with the decay of modes. In glasses Ω is determined by formulae (27), (28), (31). Thereby, the decrease of Ω at temperature lowering is connected with the reduction of the concentration of excited modes in glass and with the growth of their average distance to the impurity centre. The temperature dependence of Ω may be chiefly conditioned by modes and by interaction of a certain type or by an overall influence of various modes and interactions. In the latter case, the law $\Omega \sim T^\nu$ simply approximates the complicated dependence.

The dependence Ω^ν with $\nu = 1.85 \pm 0.2$ [25], observed in amorphous BeF_2 (Θ = 380 K) and GeO_2 (Θ = 308 K) at 8 K < T < 300 K which is close to $\Omega \sim T^2$, according to (27), can be conditioned by elastic interaction of impurity centres with quasi-local vibrations in glasses of the type a) (s = 2). Strictly speaking, such dependence (27) holds in the interval of the intermediate temperatures restricted by the conditions $\omega_\kappa^0 > 2T > \omega_4 \sim 0.1\ \Theta$. However, this law can "protract" up to temperatures several times less (i.e. up to $T \sim 0.02\ \Theta$), if a slowly-decreasing additional contribution to Ω, e.g. contribution (31), exists, which is related with two-level modes. At $T > \Theta/4$, a considerable contribution ($\sim T^2$ at $T > \Theta/2$) is made by the interaction with phonons, which results in the protraction of the law $\Omega \sim T^\nu$ with $\nu \approx 2$ into the region of high temperatures.

Another, perhaps a more probable explanation of the dependence $\Omega \sim T^\nu$ with $\nu \approx 1.8$ may be based on the account of the quadrupole-quadrupole interaction of centres with two-level modes. Thereby, according to (31), $\nu = 5/3$ (or $5(1+\mu)/3$ if $p(\Delta,\lambda) \sim \Delta^{\mu}$ with small μ) up to the lowest temperatures. In any case, it is the interaction with two-level modes (or with both types of modes), and not only with quasi-local vibrations, that determines Ω in the region of very low temperatures, lower than $\Theta/100$, and, evidently, at least partially conditions the ZPL broadening of Eu^{3+} in silicon glasses, where the law $\Omega \sim T^{1/8}$ was confirmed up to 1.6 K [23,24]. In case the dipole-quadrupole interaction with two-level modes is the basic one, the dependence T^{ν} with $\nu = 4/3$, observed in [27] (and, possibly, in [21]), should be realized, and in case of elastic and dipole interaction, the dependence $\Omega \sim T$, observed in organic glasses [26], is obtained. Naturally, the law $\Omega \sim T^{1/3}$ can also be conditioned by dipole or elastic interaction with two-level modes if $\mu \approx 0.3$, and the law $\Omega \sim T^{1/8}$ may be connected with dipole-quadrupole interaction if $\mu \approx 0.5$. Some effective increase of ν at low temperatures may also be conditioned by the contribution from the interaction with quasi-local vibrations. At $T \to 0$ the ZPL width may tend to a nonzero limit [28] if the levels s' or s belong to multiplets and radiationless transitions to lower levels of the latter are possible.

References

1 S.I. Pekar: Zh. Eksp. Teor. Fiz. 20, 510 (1950)
2 S.I. Pekar: Zh. Eksp. Teor. Fiz. 22, 641 (1952)
3 Huang Kun, A. Rhys: Proc. Roy. Soc. 204, 406 (1950)
4 M.A. Krivoglaz, S.I. Pekar: in Trudy Instituta fiziki AN SSSR, Vol. 4 (Kiev 1953) p. 37
5 M. Lax: J. Chem. Phys. 20, 1752 (1952)
6 M.A. Krivoglaz: Ph.D. Thesis, Kiev (1954)
7 R.H. Silsbee: Phys. Rev. 128, 1726 (1962)
8 D.E. McCumber: J. Math. Phys. 5, 508 (1964)
9 M.A. Krivoglaz: Fiz. Tverd. Tela 6, 1707 (1964) [English transl.: Sov. Phys. - Solid State 6, 1340 (1964)]
10 M.A. Krivoglaz: Zh. Eksp. Teor. Fiz. 48, 310 (1965) [English transl.: Sov. Phys. - JETP 21, 204 (1965) ; M.A. Krivoglaz, G.F. Levenson: Fiz. Tverd. Tela 9, 2693 (1967) [English transl.: Sov. Phys. - Solid State 9, 2114 (1968)]
11 V.V. Hizhnyakov: in Teoriya lokal'nykh tsentrov kristalla, ed. by I.J. Tehver, Trudy IFA AN ESSR, Vol. 29 (Estonian SSR Acad. Sci., Tartu 1964) p. 83
12 K.K. Rebane: Elementarnaya teoriya kolebatel'noj struktury spektrov primesnykh tsentrov kristallov (Nauka, Moskva 1968) [English transl.: Impurity Spectra of Solids (Plenum Press, New York 1970)]
13 K.V. Korsak, M.A. Krivoglaz: Fiz. Tverd. Tela 10, 2488 (1968) [English transl.: Sov. Phys. - Solid State 10, 1952 (1969)]
14 G.F. Levenson: Phys. Status Solidi b43, 739 (1971)
15 I.S. Osad'ko: Fiz. Tverd. Tela 14, 2927 (1972) [English transl.: Sov. Phys. - Solid State 14, 2252 (1972)]
16 A. Szabo: Phys. Rev. Lett. 25, 924 (1970); 27, 323 (1971)
17 R.I. Personov, E.I. Al'shitz, L.A. Bykovskaya: Pis'ma Zh. Eksp. Teor. Fiz. 15, 609 (1972) [English transl.: JETP Lett. 15, 431 (1972)]
18 A.A. Gorokhovskii, R.K. Kaarli, L.A. Rebane: Pis'ma Zh. Eksp. Teor. Fiz. 20, 474 (1974) [English transl.: JETP Lett. 20, 216 (1974)]
19 B.M. Kharlamov, R.I. Personov, L.A. Bykovskaya: Opt. Commun. 12, 191 (1974)
20 K.K. Rebane, R.A. Avarmaa: J. Photochem. 17, 311 (1981)
21 L.A. Rebane, A.A. Gorokhovskii, J.V. Kikas: Appl. Phys. B29, 235 (1982)
22 R.M.Shelby: Opt. Lett. 8, 88 (1983)
23 P.M. Selzer, D.L. Huber, D.S. Hamilton, W.M. Yen, M.J. Weber: Phys. Rev. Lett. 36, 813 (1976)

24 R.M. Macfarlane, R.M. Shelby: Opt. Commun. 45, 46 (1983)
25 J. Hegarthy, W.M. Yen: Phys. Rev. Lett. 43, 1126 (1979)
26 M. Hayes, R.P. Stout, G.J. Small: J. Chem. Phys. 74, 4266 (1981)
27 H.P.H. Thijssen, R. van den Berg, S. Völker: Chem. Phys. Lett. 97, 295 (1983)
28 J. Friedrich, H. Wolfrum, D. Haarer: J. Chem. Phys. 77, 2309 (1982)
29 A.A. Gorokhovskii, V.H. Korrovits, V.V. Palm, M.A. Trummal: Pis'ma Zh. Eksp. Teor. Fiz. 42, 249 (1985) [English transl.: JETP Lett. 42, 307 (1985)]
30 P.W. Anderson, B.I. Halperin, C.M. Varma: Phil. Mag. 25, 1 (1972)
31 W.A. Phillips: J. Low Temp. Phys. 7, 351 (1972)
32 T.L. Reineke: Solid State Commun. 32, 1103 (1979)
33 S.K. Lyo, R. Orbach: Phys. Rev. B22, 4223 (1980)
34 S.K. Lyo: Phys. Rev. Lett. 48, 688 (1982)
35 P. Reineker, H. Morawitz, K. Kassner: Phys. Rev. B29, 4546 (1984)
36 D.L. Huber, M.M. Broer, B. Golding: Phys. Rev. Lett. 52, 2281 (1984)
37 M.A. Krivoglaz: Zh. Eksp. Teor. Fiz. 88, 2171 (1985) [English transl.: Sov. Phys. - JETP 61, 1284 (1985)]
38 V.G. Karpov, M.I. Klinger, F.N. Ignat'ev: Zh. Eksp. Teor. Fiz. 84, 760 (1983) [English transl.: Sov. Phys. - JETP 57, 439 (1983)]
39 J.R. Klauder, P.W. Anderson: Phys. Rev. 125, 912 (1962)
40 V. Hizhnyakov, I. Tehver: Phys. Status Solidi b95, 65 (1979)
41 M.N. Cohen, G.S. Grest: Phys. Rev. Lett. 45, 1271 (1980)
42 M.I. Dykman, M.A. Krivoglaz: Fiz. Tverd. Tela 29, 368 (1987)

On the Theory of Stationary and Time-Dependent Zero-Phonon Lines

V.V. Hizhnyakov

Institute of Physics, Estonian SSR Acad. Sci.
SU-202400 Tartu, Estonian SSR, USSR

Abstract

The problem of the theory of zero-phonon lines (ZPL) in stationary and time-dependent (transient) optical and Mössbauer spectra of doped crystals is considered. A new method is proposed to take into account the quadratic vibronic coupling. By this method nonperturbative expressions have been found for the width and asymmetry of ZPL at low temperature and the difference of the ZPL characteristics in absorption and luminescence spectra is predicted. The inner structure of ZPL, caused by the frequency change of a slowly-decaying pseudolocal mode, is examined for arbitrary temperatures. The ZPL shape and limit width in transient optical and Mössbauer spectra are considered with the allowance made for the compensation effect, i.e. the subtraction of the natural and spectrometer resolution widths.

1. Absorption and Luminescence Spectra of Impurity Centres with Quadratic Vibronic Coupling

The shape of the optical impurity spectra of crystals and molecules is determined by the interaction of optical electrons with the vibration of nuclei [1]. In the simplest, the so-called basic model, this interaction is considered in a linear approximation with respect to phonon operators, which takes only the shifts of the equilibrium positions of vibrational oscillators on electronic transition into account. The inclusion of the change of elastic constants, which leads to the mixing up (rotation) of the system of normal vibrational coordinates (Duschinsky rotation) and to the change of their frequencies, requires the use of the quadratic vibronic interaction model. This model enables, in particular, the explanation of such well-observable effects as the lack of mirror symmetry of the luminescence and the absorption spectra as well as the temperature shift and broadening of ZPL. The latter effect is connected with the phase relaxation of the excited electronic state, that is of principle importance in the classification of resonance secondary emission, phenomena of the type of photon echo, tunnel and other transitions. The chemical aspect of the Duschinsky rotation, the change of bonds in electronic transition, is essential on considering photochemical and other reactions.

In the theory of optical spectra, usually two methods of considering the quadratic vibronic coupling are used: i) the method of density matrix of harmonic oscillators and ii) the method of T-ordered expansion of the Fourier transform of the spectrum with respect to vibronic interaction [2]. The first method, developed by KUBO and TOYOZAWA [3] enables one to obtain an exact expression for the Laplace (Fourier) transform of the spectrum. However, due to its complexity the obtained expression has not been used up to now for calculating spectra in the case of a large number of vibrational modes. The second method has been used in a large number of works (see, e.g. the book by REBANE [1], surveys by OSAD'KO [4]; the bibliography of recent years is available in the works by SKINNER and HSU, see, e.g. [5]). By this method, in particular, the laws of the temperature shift and broadening of ZPL in a crystal have been found [6-8].

The method of T-ordered expansion was further developed by LEVENSON [6], OSAD'KO [4] and others. *Levenson* and *Osad'ko*, by means of the diagram technique, deduced integral equations for functions determinig the logarithm of the Fourier transform of the spectrum. These equations have been used by *Osad'ko* to find the temperature broadening and shift of ZPL. *Skinner* and *Hsu* [5] continued the development of this theory for a number of concrete cases.

In this paper, a new nonperturbative method of considering quadratic vibronic interaction is proposed, which is based on linear multimode squeezing-type relations between phonon creation and annihilation operators in various electronic states. By this method the task of absorption spectrum has been reduced to the solution of linear equations for the Fourier amplitudes of one- and two-phonon resonance Raman scattering. A solution has been obtained for the task of the internal structure of the ZPL of the centre interacting with a pseudolocal vibration. Also, nonperturbative expansions have been found for the width and the asymmetry of the ZPL at low temperature, which differ from the corresponding formulae of *Osad'ko* and *Skinner* and *Hsu*.

The spectrum of vibronic transitions between nondegenerate electronic states of the impurity centre of a crystal in the adiabatic and Condon approximations is described by the formula [2]

$$I(\omega) = (2\pi)^{-1} \int_{-\infty}^{\infty} d\tau \, \exp(-i\omega\tau - \gamma_0|\tau|/2) \, F(\tau) \quad , \tag{1.1}$$

where γ_0 is the natural width of the excited level;

$$F(\tau) = \langle\hat{f}_\tau\rangle \equiv z_1^{-1} \, Sp(\exp(-H_1/kT) \, \hat{f}_\tau) \tag{1.2}$$

is the Fourier transform of the spectrum; $\hat{f}_\tau = \exp(-i\tau H_1) \exp(i\tau H_2)$, H_1 and H_2 are the vibrational Hamiltonians of the initial (1) and final (2) electronic states, $z_1 = Sp(\exp(-H_1/kT))$, $\hbar = 1$, and T is temperature. In the harmonic approximation

$$H_1 = \sum_i \omega_{1i}\, a^+_{1i}\, a_{1i} \;,$$
$$H_2 = \sum_j \omega_{2j}\, a^+_{2j}\, a_{2j} + \omega_0 \;, \tag{1.3}$$

where i and j number the normal coordinates of the initial (1) and final (2) electronic states, ω_{1i} and ω_{2j} are the frequencies of the corresponding normal modes, a^+_{1i} (a_{1i}) and a^+_{2j} (a_{2j}) are the operators of the creation (destruction) of quanta of these vibrations (phonons) and ω_0 is the frequency of purely electronic transition.

In the case of quadratic vibronic interaction

$$V = H_2 - H_1 = V_0 + (aq) + \frac{1}{2}(qbq) \;, \tag{1.4}$$

where the vector a and the tensor b describe the changes of the forces (a) and elastic constants (b) on electronic transition; their dimension n is determined by the number (usually small) of elastic constants changing on electronic transition;

$$q = \sum_i e_{1i}x_i = \sum_j e_{2j}y_j + q_0 \tag{1.5}$$

is the vector of the configurational coordinates of the centre; $x_i = (a^+_{1i} + a_{1i})\times(2\omega_{1i})^{-1/2}$ and $y_j = (a^+_{2j} + a_{2j})(2\omega_{2j})^{-1/2}$ are the normal coordinates in the initial and final electronic states related by the orthogonal transformation

$$y_j = y_{0j} + \sum_j c_{ij}x_i \tag{1.6}$$

(Duschinsky rotation); c_{ij} is the rotation matrix equal to [7]

$$c_{ij} = (e_{1i}be_{2j})/(\omega^2_{2j} - \omega^2_{1i}) \;, \tag{1.7}$$

$y_{0j} = (ae_{2j})\omega^{-2}_{2j}$; $q_0 = -\sum_j e_{2j}y_{0j}$; e_{1i} and e_{2j} are the components of the vector q in the space of the coordinates x_i and $y_j - y_{0j}$ which satisfy the condition

$$e_{1i} = \sum_j c_{ij}e_{2j} \;, \quad e_{2j} = \sum_i c_{ij}e_{1i} \;, \quad \sum_i e^2_{1i} = \sum_j e^2_{2j} = 1 \;. \tag{1.8}$$

In the present model, the phonon creation and destruction operators in different electronic states are related by the multimode squeezing-type linear transformations:

$$a^+_{1i} = \xi_{1i} + \frac{1}{2}\sum_j \frac{c_{ij}}{\sqrt{\omega_{1i}\omega_{2j}}}\left[(\omega_{1i} + \omega_{2j})a^+_{2j} + (\omega_{1i} - \omega_{2j})a_{2j}\right] \;, \tag{1.9a}$$

$$a^+_{2j} = \xi_{2j} + \frac{1}{2}\sum_\kappa \frac{c_{\kappa j}}{\sqrt{\omega_{1\kappa}\omega_{2j}}}\left[(\omega_{1\kappa} + \omega_{2j})a^+_{1\kappa} + (\omega_{2j} - \omega_{1\kappa})a_{1\kappa}\right] \;, \tag{1.9b}$$

where $\xi_{1i} = (bq_0 - a)e_{1i}(2\omega_{1i}^3)^{-1/2}$, $\xi_{2j} = ae_{2j}(2\omega_{2j}^3)^{-1/2}$; relations for a_{1i} and a_{2j} are obtained by conjugation from those presented. Formulae (1.9) result from the transformation (1.6) for the components of a covariant vector and from an analogous transformation $\partial/\partial y_j = \sum_i c_{ij}\partial/\partial x_i$ for the components of a contravariant vector.

Let us regard a derivative of the Fourier transform:

$$-i\frac{dF(\tau)}{d\tau} = \langle\hat{f}_\tau V\rangle = \sum_i (ae_{1i})(1 + e^{-\lambda_i})\, P_i(\tau) + \frac{1}{2}\sum_{ii'}(a_{1i}b\, e_{1i'}) \times (1 + e^{-\lambda_i - \lambda_{i'}})\, R_{ii'}(\tau) + 2S_{ii'}(\tau) + \frac{1}{2\omega_{1i}}\delta_{ii'} \quad , \tag{1.10}$$

where $\lambda_i = \omega_{1i}(i\tau + 1/kT)$,

$$P_i(\tau) = \langle\hat{f}_\tau\, a^+_{1i}\rangle\, (2\omega_{1i})^{-1/2} \quad , \tag{1.11}$$

$$R_{ii'}(\tau) = \frac{1}{2}\langle\hat{f}_\tau\, a^+_{1i}a^+_{1i'}\rangle\, (\omega_{1i}\omega_{1i'})^{-1/2} \quad , \tag{1.12}$$

$$S_{ii'}(\tau) = \frac{1}{2}\langle\hat{f}_\tau\, a^+_{1i}a_{1i'}\rangle\, (\omega_{1i}\omega_{1i'})^{-1/2} \quad , \tag{1.13}$$

On obtaining formula (1.10) we took into consideration that

$$\langle\hat{f}_\tau\, a_{1i}\rangle = e^{-\lambda_i}\langle\hat{f}_\tau\, a^+_{1i}\rangle \quad , \tag{1.14}$$

$$\langle\hat{f}_\tau\, a_{1i}\, a_{1i'}\rangle = e^{-\lambda_i - \lambda_{i'}}\langle\hat{f}_\tau\, a^+_{1i}\, a^+_{1i'}\rangle \quad . \tag{1.14a}$$

It is of interest to note that the functions $\Theta(\tau)\langle\hat{f}_\tau\, a^+_{1i}\rangle$ and $\Theta(\tau)\langle\hat{f}_\tau\, a_{1i}\rangle$ determine the Fourier amplitudes of the first-order Stokes and anti-Stokes resonance Raman scattering (RRS), while the functions $\Theta(\tau)\langle\hat{f}_\tau\, a^+_{1i}\, a^+_{1i'}\rangle$, $\Theta(\tau)\langle\hat{f}_\tau\, a^+_{1i}\, a_{1i'}\rangle$ and $\Theta(\tau)\langle\hat{f}_\tau\, a_{1i}a_{1i'}\rangle$ define the Fourier amplitudes of the second-order RRS [8] ($\Theta(\tau)$ is the Heaviside step function).

From formulae (1.9b) and (1.14) it follows that

$$\langle\hat{f}_\tau\, a^+_{2j}\rangle = \xi_{2j}F(\tau) + \sqrt{\omega_{2j}/2}\sum_\kappa c_{\kappa j}\left[1 + \frac{\omega_{1\kappa}}{\omega_{2j}} + \left(1 - \frac{\omega_{1\kappa}}{\omega_{2j}}\right) e^{-\lambda_\kappa}\right]\langle\hat{f}_\tau\, a^+_{1\kappa}\rangle \quad . \tag{1.15}$$

Let us replace in the left-hand side of (1.15) the operators $\exp(i\tau H_2)$ and a^+_{2j} according to $\exp(i\tau H_2)a^+_{2j} = a^+_{2j}\exp(i\tau(H_2 + \omega_{2j}))$, express a^+_{2j} via $a_{1\kappa}$ and $a^+_{1\kappa}$ (see (1.9b)) and consider that

$$<e^{-i\tau H_1} a_{1\kappa} e^{i\tau H_2}> = <\hat{f}_\tau a^+_{1\kappa}> \quad ,$$

$$<e^{-i\tau H_1} a^+_{1\kappa} e^{i\tau H_2}> = e^{-\lambda_\kappa} <\hat{f}_\tau a^+_{1\kappa}> \quad .$$

We obtain

$$<\hat{f}_\tau a^+_{2j}> = e^{i\tau\omega_{2j}} \{\xi_{2_j} F(\tau) + \frac{1}{2} \sum_\kappa c_{\kappa j} (\omega_{1\kappa}\omega_{2j})^{-1/2} \} [(\omega_{2j} - \omega_{1\kappa})$$
$$+ (\omega_{2j} + \omega_{1\kappa}) e^{-\lambda_\kappa}] <\hat{f}_\tau a^+_{1\kappa}> \quad . \qquad (1.15a)$$

Multiplying the right-hand sides of (1.15) and (1.15a) by $c_{ij}(\omega_{2j})^{-1/2}$ and summing over j with account of the orthogonality condition

$$\sum_j c_{ij} c_{\kappa j} = \delta_{i\kappa} \quad ,$$

we obtain for the normalized RRS amplitude

$$p_i = F_i^{-1} (2\omega_{1i})^{-1/2} <\hat{f}_\tau a^+_{1i}> \qquad (1.16)$$

the equation

$$p_i = A_i + \sum_\kappa (B_{i\kappa} + \bar{B}_{i\kappa}) p_\kappa \quad . \qquad (1.17)$$

Here

$$A_i = \frac{1}{2} \sum_j (ae_{2j}) c_{ij} \omega_{2j}^{-2} (1 - e^{i\tau\omega_{2j}}) \quad , \qquad (1.18)$$

$$B_{i\kappa} = \frac{1}{2} \sum_j c_{ij} c_{\kappa j} [(1 + \frac{\omega_{1\kappa}}{\omega_{2j}}) e^{i\tau\omega_{2j} - \lambda_\kappa} + 1 - \frac{\omega_{1\kappa}}{\omega_{2j}}] \quad , \qquad (1.19)$$

$$\bar{B}_{i\kappa} = \frac{1}{2} \sum_j c_{ij} c_{\kappa j} (1 - \frac{\omega_{1\kappa}}{\omega_{2j}}) (e^{i\tau\omega_{2j}} - e^{-\lambda_\kappa}) \quad . \qquad (1.20)$$

An expression equivalent to (1.17) at T = 0 was obtained earlier in [9] when examining the effect of the Duschinsky rotation upon the first-order RRS.

Considering analogously the correlators $<\hat{f}_\tau a^+_{2j} a^+_{1i'}>$ and $<\hat{f}_\tau a_{2j} a^+_{1i'}>$ and taking the following relations into account

$$<e^{-i\tau H_1} a^+_{1\kappa} e^{i\tau H_2} a^+_{1i'}> = e^{-\lambda_\kappa} <\hat{f}_\tau a^+_{1\kappa} a^+_{1i'}> \quad ,$$

$$<e^{-i\tau H_1} a_{1\kappa} e^{i\tau H_2} a^+_{1i'}> = e^{-\lambda_{i'}} <\hat{f}_\tau a^+_{1\kappa} a_{1i'}> \quad ,$$

we obtain for the matrices

$$r_{ii'} = F^{-1}\langle \hat{f}_\tau \, a^+_{1i} \, a^+_{1i'} \rangle \, (\omega_{1i}\omega_{1i'})^{-1/2} \tag{1.21}$$

and

$$s_{ii'} = F^{-1}\langle \hat{f}_\tau \, a^+_{1i} \, a_{1i'} \rangle \, (\omega_{1i}\omega_{1i'})^{-1/2} \tag{1.22}$$

the equations

$$r_{ii'} = 2A_i p_{i'} + D_{ii'} + \sum_\kappa (B_{i\kappa} r_{\kappa i'} + \bar{B}_{i\kappa} s_{\kappa i'} \, e^{\lambda_{i'}}) \quad , \tag{1.23}$$

$$s_{ii'} = 2A_i p_{i'} \, e^{-\lambda_{i'}} + \bar{D}_{ii'} + \sum_\kappa (B_{i\kappa} s_{\kappa i'} + \bar{B}_{i\kappa} r_{\kappa i'} \, e^{-\lambda_{i'}}) \quad , \tag{1.24}$$

where

$$D_{ii'} = \frac{1}{2} \sum_j c_{ij} c_{i'j} (\omega_{2j}^{-1} - \omega_{1i'}^{-1}) \quad , \tag{1.25}$$

$$\bar{D}_{ii'} = \frac{1}{2} \sum_j c_{ij} c_{i'j} (\omega_{2j}^{-1} - \omega_{1i'}^{-1}) \, e^{i\tau\omega_{2j} - \lambda_{i'}} \quad . \tag{1.26}$$

Based on the formula obtained, (1.10)-(1.13) and (1.16)-(1.26), the Fourier transform of the spectrum takes the form

$$F(\tau) = \exp(i \int_0^\tau d\tau' \, L(\tau')) \quad , \tag{1.27}$$

where

$$L(\tau) = \sum_i (ae_{1i}) p_i (1 + e^{-\lambda_{i'}}) + \frac{1}{4} \sum_{ii'} (e_{1i} b e_{1i'})$$

$$\times \, [(1 + e^{-\lambda_i - \lambda_{i'}}) r_{ii'} + 2s_{ii'} + \omega_{1i}^{-1} \delta_{ii'}] \quad . \tag{1.28}$$

Thus, the task of the Fourier transformation of the spectrum reduces to the solution of linear inhomogeneous matrix equations (1.17), (1.23) and (1.24). These equations can be solved exactly (in case of not large numbers of vibrational degrees of freedom N) or approximately, for example, by the iteration method (in case of arbitrary N).

Let us note some peculiarities of the expression obtained for the Fourier transform of the spectrum.

The Fourier transform contains only positive powers of the exponents $\exp(i\tau\omega_{2j})$ and only negative powers of the exponents $\exp(i\tau\omega_{1i})$, whereat the latter exist only in the combination $\exp(-\omega_{1i}(i\tau + 1/kT))$. Therefore, as it is natural from physical considerations, the spectrum consists of the lines corresponding to the transitions with the creation of phonons in the final electronic state (with frequencies ω_{2j}) and with their annihilation in the initial electronic state (with ω_{1j}). The intensity of the line, corresponding to annihilation of the phonon of the frequency ω_{1i}, is proportional to the mean number of such phonons $\bar{n}_i$.

At zero temperature (T = 0) the spectrum contains only ZPL and the lines corresponding to the transitions with the creation of phonons in the final electronic state. The formulae obtained above become simplified in this case. In particular, at weak and intermediate vibronic interaction, when the condition $|b|\bar{\omega}^{-2} \ll 1$ ($\bar{\omega}$ is the average phonon frequency) is fulfilled, one may take $p_i \simeq A_i + \sum_{\kappa} (B_{i\kappa} + \bar{B}_{i\kappa})A_{\kappa}$, $r_{ii'} \simeq 2A_iA_{i'}$. In this approximation

$$L(\tau;\ T=0) \simeq (aq_{\tau}) + \frac{1}{2}(q_{\tau}bq_{\tau}) + \frac{1}{2}(a\phi_{\tau}a) \quad , \qquad (1.29)$$

where

$$q_{\tau} = a\pi^{-1} \int_0^{\infty} d\omega\ \omega^{-1}(e^{i\omega\tau} - 1)\ \mathrm{Im}\{G_2(\omega)\} \quad , \qquad (1.30)$$

$$\phi_{\tau} = \pi^{-2} \iint_0^{\infty} d\omega\ d\omega' \frac{\mathrm{Im}\{G_2(\omega)b\}\,\mathrm{Im}\{G_2(\omega')\}}{(\omega + \omega')\omega'} (1 + e^{i\omega\tau})(1 - e^{i\omega'\tau}) \quad . \qquad (1.31)$$

Here the dynamical Green's function of the final electronic state has been introduced

$$G_2(\omega) = \sum_j e_{2j}e_{2j}(\omega^2 - \omega_{2j}^2 - i\varepsilon\omega)^{-1} \quad , \qquad \varepsilon \to 0 \quad . \qquad (1.32a)$$

This Green's function is associated with the dynamical Green's function of the initial state

$$G_1(\omega) = \sum_i e_{1i}e_{1i}(\omega^2 - \omega_{1i}^2 - i\varepsilon\omega)^{-1} \quad , \qquad \varepsilon \to 0 \qquad (1.32b)$$

by the Lifshitz formula

$$G_2(\omega) = (I + bG_1(\omega))^{-1}\ G_1(\omega) \quad .$$

Note that the absorption and luminescence spectra in approximation (1.29) do not possess any mirror symmetry due to the difference of the dynamical Green's functions of the ground and excited electronic states; besides, the parameters of a linear vibronic coupling for absorption and luminescence differ not only in the sign but also in the absolute value if $b \neq 0$.

One of the important temperature effects in spectra is the broadening of ZPL. It is the result of the transitions with the creation and annihilation of phonons of close frequencies [10,11], described in (1.28) by the items proportional to $s_{ii'}(\tau)$, which give the finite contribution to the asymptotics of $L(\tau)$ at $\tau\to\infty$ (see (1.28)); only these items contain terms $\sim[1 - \exp(i\tau(\omega_{2j} - \omega_{1\kappa}))]/(\omega_{2j} - \omega_{1\kappa})$, $\omega_{1\kappa}\to\omega_{2j}$, which do not decay at $\tau\to\infty$. Thereby,

$$s_{ii'} = \bar{n}_i\delta_{ii'} + \sigma_{ii'} \quad , \tag{1.33}$$

where $\bar{n}_i = (\exp(\omega_i/\kappa T) - 1)^{-1}$,

$$\begin{aligned} \sigma_{ii'} &= (\bar{n}_i + 1) \sum_\kappa d_{i\kappa}(\bar{n}_{i'}\delta_{i'\kappa} + e^{-\omega_{1\kappa}/\kappa T}\sigma_{\kappa i'}) \quad , \\ d_{i\kappa} &= \sum_j \frac{c_{ij}(e_{1\kappa}be_{2j})\omega_{1i}^{1/2}}{2\omega_{2j}\omega_{1\kappa}^{1/2}(\omega_{2j} - \omega_{1\kappa})}\left(e^{i\tau(\omega_{2j}-\omega_{1\kappa})} - 1\right) \quad . \end{aligned} \tag{1.34}$$

At low temperatures and an arbitrary b (as well as for arbitrary temperature and small $|b|$) the items $\sim\exp(-\omega_{1\kappa}/\kappa T)\sigma_{\kappa i'}$ can be neglected. Then we get

$$F(\tau) \sim \exp(-\gamma|\tau| + i\alpha \,\mathrm{sign}\, \tau) \quad , \qquad \tau \to \infty \quad . \tag{1.35}$$

Here

$$\gamma = \mathrm{Re} \int_0^\infty d\tau \, (bg_2(\tau)\, bg_1(-\tau)) \quad , \tag{1.36}$$

$$\alpha = \mathrm{Im} \iint_0^\infty d\tau \, d\tau' \quad (bg_2(\tau+\tau')\, bg_1(-\tau-\tau')) \quad , \tag{1.37}$$

where

$$g_1(\tau) = \pi^{-1} \int_0^\infty d\omega \, e^{i\omega\tau} \, \bar{n}_\omega \, \mathrm{Im}\{G_1(\omega)\} \; ,$$

$$g_2(\tau) = \pi^{-1} \int_0^\infty d\omega \, e^{i\omega\tau} \, (\bar{n}_\omega + 1) \, \mathrm{Im}\{G_2(\omega)\} \; ,$$

$\bar{n}_\omega = (\exp(\omega/\kappa T) - 1)^{-1}$. The parameter γ determines the broadening of ZPL and the rate of the pure dephasing of electronic excitation and the parameter α, the asymmetry of ZPL. These parameters are, in general, different for absorption and luminescence (the difference decreases if $|b|\to 0$ and(or) $T\to 0$).

The above-mentioned conclusion of this theory about the difference of ZPL characteristics in absorption and luminescence is opposite to the one of the nonperturbative theory of *Osad'ko* [4] and *Skinner* and *Hsu* [5]. According to their theory the broadening of ZPL at low temperatures is determined by the following spectral integral (ZPL asymmetry was not considered in [4,5]):

$$\gamma_{0s} = \pi^{-1} b^2 \int_0^\infty d\omega\, \bar{n}_\omega(\bar{n}_\omega + 1)\, \mathrm{Im}\{G_1(\omega)\}\mathrm{Im}\{G_2(\omega)\}$$

(b is taken one-dimensional, (n = 1)) . In this expression, the local phonon densities $\mathrm{Im}\{G_{1,2}(\omega)\}$ of the initial and final electronic states enter symmetrically, which results in the equality of γ in absorption and luminescence. Our expression (1.36) has no such symmetry.

To elucidate the difference of our and *Osad'ko's* results let us also present (1.36) in the spectral form (n = 1):

$$\gamma = \frac{1}{2} b^2 \sum_i e^2_{1i}\, \omega^{-1}_{1i}\, \bar{n}_i(\bar{n}_i + 1)\, \mathrm{Im}\{G_2(\omega_{1i})\} \tag{1.36a}$$

$$= \pi^2 b^2\, \mathrm{Im}\{\oint \bar{n}_\omega(\bar{n}_\omega + 1)\, G_2(\omega)\}\, \mathrm{Im}\{\bar{G}_1(\omega)\} \ . \tag{1.36b}$$

Here

$$\bar{G}_1(\omega) = \frac{1}{2} \sum_i e^2_{1i}\, \omega^{-1}_{1i}(\omega - \omega_{1i} - i\varepsilon)^{-1} \quad , \qquad \varepsilon \to 0;$$

integration in (1.36b) is performed over the contour comparising the lower complex half-plane, where $\bar{G}_1(\omega)$ has poles, while $G_2(\omega)$ has not. We can see that in our expression (1.36b) only the poles of $G_1(\omega)$ with $\mathrm{Re}\{\omega\} > 0$ give a contribution to γ, while according to *Osad'ko's* expression the poles of both Green's functions, $G_2(\omega)$ and $G_1(\omega)$, with $\mathrm{Re}\{\omega\} > 0$, contribute to γ_{0s}.

In the case of small $|b|$ one can take $G_2(\omega) \simeq G_1(\omega)$ and replace in (1.36a) the sum over i by the integral over vibrational frequencies. Then γ and γ_{0s} coincide with each other and with the corresponding *McCumber's* expression [10]. But for arbitrary b the replacing is incorrect because of the change of phonon frequencies in electronic transition.

At very low temperatures expressions (1.36) and (1.37) give $\gamma \sim T^7$ and $\alpha \sim T^6$ (for arbitrary b, compatible with the lattice stability condition in the excited state $b > -(\mathrm{Re}\{G_1(0)\})^{-1}$; here we take into account that in crystals $\mathrm{Im}\{G_{1,2}(\omega)\} \sim \omega^3$, $\omega \to 0$, as a result of which $g_2(\tau) b g_1(-\tau) \sim \tau^{-8}(1 + i/\kappa T)$, $\tau \geq 1/\kappa T \to \infty$). These results generalize analogous results for γ [10] and α [12] proved eralier for small $|b|$.

Let us examine now the structure of ZPL caused by the transitions from various levels of a slowly-damping local or pseudolocal mode (PLM) whose frequency changes on electronic transition. Such structure has earlier been investigated in [13,14] (see also [1]). In this case, for the model with one elastic constant changing on the electronic transition

$$G_{1,2}(\omega) = (\omega^2 - \omega^2_{1,2} - 2i\omega\Gamma)^{-1} \quad , \tag{1.38}$$

where ω_1 and $\omega_2 = \omega_1 + \Delta = (\omega_1 + b)^{1/2}$ are the frequencies of PLM in the initial and final electronic states, Δ, their difference, Γ, the decay constant of PLM. Thus, PLM represent a phonon packet. The ZPL structure of interest is due to the transitions with the creation and destruction of these phonons and, therefore, is described by the items in (1.28) proportional to $s_{ii'}(\tau)$.

Let us assume that the condition $\Gamma, |\Delta| << \omega_1$ is fulfilled. In this case in equation (1.24) for $s_{ii'}$, the terms $\sim p_i$ and $\sim r_{ii'}$ can be omitted and in the coefficients $\bar{D}_{ii'}$, the terms $\sim (1 - \omega_{1i'}/\omega_{2j})$ neglected. Besides, within the boundaries of small and large Γ (in comparison with $|\Delta|$) the approximation

$$s_{ii'} \simeq e_{1i} e_{1i'} s \tag{1.39}$$

may be used, where s satisfies the equation

$$s(\tau) = e^{-\omega_1/\kappa T} R(\tau) (1 + s(\tau)) \quad , \tag{1.40}$$

where

$$R(\tau) = \sum_{ij} e_{1i} e_{2j} c_{ij} \, e^{i(\omega_{2j} - \omega_{1i})\tau}$$
$$\simeq (\Delta + i\Gamma)^{-1} (\Delta \, e^{i\Delta\tau - \Gamma\tau} + i\Gamma) \quad , \qquad \tau \geq 0 \quad . \tag{1.41}$$

By iterating equation (1.24) with (1.39) as the zero-th approximation it follows that equation (1.40) is valid when the relation

$$\frac{\Delta\Gamma \exp(-2\omega_1/\kappa T)}{\Delta^2 + \Gamma^2} << 1 \tag{1.42}$$

takes place.

The solution of (1.40) is the following:

$$s(\tau) = \frac{\zeta \exp(-\omega_1/\kappa T) \left[\exp(i\tau(\Delta + i\Gamma) + i\Gamma/\Delta\right]}{1 - \zeta \exp(-\omega_1/\kappa T) \exp(i\tau(\Delta + i\Gamma))} \quad , \tag{1.43}$$

where

$$\zeta = (1 + i\Gamma\Delta^{-1} (1 - e^{-\omega_1/\kappa T}))^{-1} \quad . \tag{1.44}$$

Substituting (1.43) into (1.28) and (1.27) and integrating over τ', we obtain the following expression for the factor of the Fourier transform describing ZPL

$$F^{(0)}(\tau) = \exp\{-\beta - \tau(\gamma + i\delta) - \zeta \ln(1 - \zeta \, e^{-\omega_1/\kappa T} \, e^{i\tau(\Delta + i\Gamma)})\} \quad , \tag{1.45}$$

where

$$\beta = \zeta \ln(1 - \zeta e^{-\omega_1/\kappa T}) \quad , \tag{1.46}$$

$$\delta = \frac{1}{2}\Delta + \Delta\Gamma^2\bar{n}/[\Delta^2(\bar{n}+1)^2 + \Gamma^2] \; , \tag{1.47}$$

$$\gamma = 2\Delta^2\Gamma\bar{n}(\bar{n}+1)/[\Delta^2(\bar{n}+1)^2 + \Gamma^2] \; , \tag{1.48}$$

$\tau \geq 0$; the Fourier transform at $\tau < 0$ is found from the condition $F(-\tau) = F^*(\tau)$. ZPL is split into components at $\Gamma \ll |\Delta|$, when $\zeta \simeq 1$. Hereby

$$F^{(o)}(\tau) \simeq (1 - e^{-\omega_1/\kappa T})\, e^{i\delta\tau - \gamma\tau}\{e^{i\alpha_o}$$

$$+ \sum_{m=o}^{\infty} \exp[(-m\omega_1/\kappa T) + im\tau(\Delta + i\Gamma) + i\alpha_m]\} \quad , \tag{1.49}$$

where

$$\alpha_o = \frac{\Gamma}{\Delta(\bar{n}+1)}(\bar{n} + \ln(\bar{n}+1)) \quad ,$$

$$\alpha_m = \alpha_o - \frac{\Gamma}{\Delta(\bar{n}+1)}(m + \sum_{\kappa=1}^{m} \kappa^{-1}) \quad .$$

Indeed, the spectrum described by Fourier transform (1.49) consists of a series of equidistant lines that broaden linearly and decay exponentially with the number of line. The relative intensity of the m-th line is determined by the Boltzmann population of the m-th level of pseudolocal vibration that is equal to $(1 - \exp(-\omega_1/\kappa T)) \times \exp(-m\omega_1/\kappa T)$; the distance between the neighbouring lines in the spectrum equals Δ and their width is equal to $\gamma_0 + \gamma + 2\Gamma m$, while the broadening $2\Gamma m$ is the Weisskopf-Wigner broadening of the transition between the levels of the widths Γm. Thus, the Fourier transform $F^{(o)}(\tau)$ really describes the inner structure of a ZPL, which results from transitions between different levels of pseudolocal vibration. Note that all ZPL components are asymmetrical. The asymmetry of the main component ($m = 0$) is determined by the parameter Im β; the parameter γ determines the temperature broadening of this component and δ, its temperature shift.

At low temperatures ($\bar{n} \ll 1$), $\gamma = 2\Gamma\Delta^2\bar{n}(\Delta^2 + \Gamma^2)^{-1}$. In the limit $\Gamma \gg |\Delta|$, this formula coincides with the *Krivoglaz* formula [6] $\gamma \simeq 2\Delta^2\bar{n}/\Gamma$ for the width of ZPL. At $\Gamma \ll |\Delta|$ and temperatures $T \gg \Gamma/\kappa$ the width of the main component of ZPL equals $\gamma \simeq 2\Gamma \exp(-\omega_1/\kappa T)$. In this case at low temperatures $\gamma \ll 2\Gamma$, and at high temperatures $\gamma \simeq 2\Gamma$.

In case of a few PLM, the Fourier transform of ZPL should be calculated by the product of factors (1.45), each of which takes into account one PLM. If there are many PLM, the components of the inner structure of ZPL will overlap, and the shape of ZPL as a whole will turn, with the rise of temperature, to a Gaussian with quadratic dispersion $\sum_{\kappa} \Delta_{\kappa}^2 \bar{n}_{\kappa}(\bar{n}_{\kappa} + 1)$ [6,7] (here κ is the PLM index).

2. Time-Dependent ZPL Spectrum; Compensation Effect

One of the momentous problems in the theory of time-dependent (transient) spectra is the limitingly obtainable value of the linewidth in them. The solution of the problem, in its turn, depends on what time should be used in the uncertainty principle for light frequency and time. Below the following solution is given: in the time-dependent spontaneous emission spectra measured by a regular scheme (with the help of a spectrometer and a photon counter), the limiting spectral resolution is determined by the reverse value of the time interval t between the excitation of the light source and the recording of the triggered photon by the counter. Thereby the limiting linewidth t^{-1} can be obtained only in case of the correctly-adjusted spectrometer resolution η. In particular, in a time-dependent spectrum of resonant fluorescence (luminescence), the linewidth depends (apart from t and other parameters) on $|\gamma_0 - \eta|$, while in time-dependent spectra of absorption and scattering, it depends on $|\gamma_0 - \eta_0|$ and $|\eta - \eta_0|$, respectively, where γ is the natural linewidth and η_0, the spectral linewidth of the exciting light pulse. As a result, lines with $\sim t^{-1}$ linewidth can be observed at $\eta = \gamma$ (fluorescence), $\eta_0 = \gamma_0$ (absorption) and at $\eta = \eta_0$ (scattering); in these cases, η, η_0 and γ_0 are compensated and they fall out from the linewidth formulae.

This effect of the compensation of the spectrometer resolution has been considered in [15,16] within the frames of the quantum theory of resonant secondary radiation, and also in [17-20] (absorption) and [21] (resonance fluorescence). Note also paper [22], which showed that in a transient spectrum of the transmission of Mössbauer γ-quanta through the resonant absorber, a minimum with the halfwidth t^{-1} can also be observed.

Narrow lines with the widths $\sim t^{-1}$ have also been observed in time-dependent excitation spectra [23] (see also [24]). These lines are conditioned by induced oscillations of the optical transition moment, not by the spontaneous ones considered here and in [15,16]. In the experiment [24], no spectral device has been used; there is no compensation effect of the spectrometer resolution either.

Below it will be shown that in time-dependent (transient) spectra of the systems with large inhomogeneous or Doppler broadening, lines with the width t^{-1} can also be observed, while here a double compensation takes place: γ_0 and η as well as η and η_0 are compensated. This allows the compensation effect to be observed on inhomogeneously strongly-broadened ZPL in low-temperature spectra of impurity crystals and on Doppler-broadened spectral lines of low-pressure gases. The consideration is based on a classical theory of emission.

Let us define the transient emission spectrum as the dependence of the counting rate of the photons passed through the spectrometer by a spectrally nonsensitive detector on the time t and on the spectrometer transmission frequency Ω at the assigned spectral resolution η [15,16]:

$$I(t,\Omega) = I(t,\Omega,\eta) \sim |E(t,\Omega,\eta)|^2 \quad , \tag{2.1}$$

where $E(t,\Omega,\eta) \equiv E(t)$ is the intensity of the light wave passing through the spectrometer at the time moment t.

As a result of the filtration of waves by the spectrometer, the form of the passed-through signal will be changed:

$$E(\omega) = E^1(\omega)\, C(\omega) \quad , \tag{2.2}$$

$E^1(\omega)$ is the Fourier component of the electric field of the incident emission and $E(\omega)$, that of the emission passed through the spectrometer, $|C(\omega)|^2$ is the spectrometer filtration function defining the law of transforming the intensities of monochromatic waves by spectrometer. From formula (2.2) it follows that the incident and passed-through fields are connected by an integral relation

$$E(t) = \frac{1}{2\pi} \int_{-\infty}^{\infty} dt_1\, E^1(t_1)\, C(t-t_1) \quad , \tag{2.3}$$

where C(t) is the field transmission function (the Fourier transform of the function $C(\omega)$). At $t < 0$, because of the causality principle, C(t) equals zero: $C(t) = \Theta(t)C(t)$. Consequently, its Fourier transform $C(\omega)$ can have poles only in the upper part of the complex plane. In particular this feature is satisfied by the following function:

$$C(\omega) = iC_0(\Omega - \omega + i\eta)^{-1} \tag{2.4}$$

(C_0 is a constant), which describes the Fabry-Perot etalon. In this case

$$C(t) = 2\pi C_0\, \Theta(t)\, e^{i\Omega t - \eta t} \tag{2.5}$$

and

$$E(t) = C_0 \int_{-\infty}^{t} dt_1\, e^{i\Omega(t-t_1) - \eta(t-t_1)}\, E^1(t) \quad . \tag{2.6}$$

Let us regard a classical radiation field emitted by a single atom (nucleus) after its instantaneous excitation at the time moment t = 0. This field, evidently, is as follows:

$$E^1(t) = \varepsilon_0\, \Theta(t)\, e^{i\omega_0 t - \gamma_0 t} \quad , \tag{2.7}$$

where ε_0 is the amplitude and ω_0, the frequency of the electronic (nuclear) transitions. A substitution of formula (2.7) into (2.6) gives

$$E(t) = i\varepsilon_0 C_0 e^{i\omega_0 t-\gamma_0 t}\left[\frac{1 - e^{i(\Omega-\omega_0)t - (\gamma_0-\eta)t}}{\Omega_0 - \omega_0 - i(\gamma_0 - \eta)}\right] .$$

Therefore, the time-dependent emission (fluorescence) spectrum equals [15,20]

$$I(t,\Omega) = |\varepsilon_0 C_0|^2 \frac{e^{-2\gamma_0 t} + e^{-2\eta t} - 2e^{-(\gamma_0+\eta)t}\cos(\Omega - \omega_0)t}{(\Omega - \omega_0)^2 + (\gamma_0 - \eta)^2} . \qquad (2.8)$$

From this formula it follows that the minimum linewidth in the spectrum is obtained at $\gamma_0 = \eta$, when $I \sim x^{-2}(1 - \cos xt)$, where $x = \Omega - \omega_0$; in this case the spectrum oscillates, while the width of the main central maximum equals t^{-1}. Thus, in a time-dependent spectrum the minimum linewidth really equals t^{-1}, whereby an effect of the compensation of γ_0 and η takes place.

If the exciting pulse has the finite duration, then the radiation field of an atom is described by the formula

$$E'(t_1) = \int_{-\infty}^{t_1} dt_2\, e^{i\omega_0(t_1 - t_2) - \gamma_0(t_1 - t_2)} E_1(t_2) , \qquad (2.9)$$

where $E_1(t_2)$ determines the time dependence of the electric field of the exciting light wave. At such excitation the transient spectrum of resonant secondary emission (RSE) is described by the following formula:

$$I \sim \left| \int_{-\infty}^{t} dt_1 \int_{-\infty}^{t_1} dt_2\, C(t - t_1)\, e^{i\omega_0(t_1 - t_2) - \gamma_0(t_1 - t_2)} E_1(t_2)\right|^2 . \qquad (2.10)$$

If the excitation is carried out through a monochromator with a Lorentzian filtration function, then

$$E_1(t_2) \sim \Theta(t_2)\, e^{i\Omega_0 t_2 - \eta_0 t_2} , \qquad (2.11)$$

where Ω_0 is the average excitation frequency and η_0, its spectral width. In this case the radiation field is of the following form:

$$E'(t_1) \sim \frac{\Theta(t_1)\left(e^{i\omega_0 t_1 - \gamma_0 t_1} - e^{i\Omega_0 t_1 - \eta_0 t_1}\right)}{\eta_0 - \gamma_0 + i(\omega_0 - \Omega_0)} . \qquad (2.12)$$

Substituting this formula into formula (2.10), after integrating with the allowance made for (2.5), one obtains

$$I \sim \frac{e^{-2\eta t}}{x_o^2 + \bar{\gamma}_o^2} \left[\frac{1 + e^{-2\bar{\gamma} t} - 2e^{-\bar{\gamma} t} \cos xt}{x^2 + \bar{\gamma}^2} + \frac{1 + e^{2\bar{\eta} t} - 2e^{\bar{\eta} t} \cos yt}{y^2 + \bar{\eta}^2}\right.$$

$$\left. - 2\mathrm{Re}\left\{\frac{(1 - e^{ixt-\bar{\gamma} t})(1 - e^{iyt+\bar{\eta} t})}{(x + i\bar{\gamma})(y - i\bar{\eta})}\right\}\right] , \qquad (2.13)$$

where $x_0 = \Omega_0 - \omega_0$, $x = \Omega - \omega_0$, $y = \Omega_0 - \Omega$, $\bar{\gamma}_0 = \gamma_0 - \eta_0$, $\bar{\gamma} = \gamma_0 - \eta$, $\bar{\eta} = \eta - \eta_0$. In this formula, the first term in the square brackets describes the luminescence-type RSE, the second one, the scattering-type RSE, and the third one, their interference. The luminescence linewidth is determined by the parameter $|\gamma_0 - \eta|$ and t^{-1}, the scattering linewidth, by $|\eta - \eta_0|$ and t^{-1}. Consequently, in the luminescence line η and γ_0, in the scattering line η and η_0, are compensated.

In doped crystals, at low temperatures ZPL are inhomogeneously broadened. The effect of the broadening on the time-dependent RSE spectrum can be found by integrating $I(t,\Omega)$ (see formula (2.13)) over ω_0 with the corresponding distribution function $f(\omega_0)$. Analogously, the Doppler broadening of atomic lines can be taken into account in gases. As a rule, the width (γ_2) of the distribution $f(\omega_0)$ exceeds γ_0 by several orders ($\gamma_2 \sim 10^3$-$10^5\ \gamma_0$). Therefore, it can be taken that $f(\omega_0)$ = const. In this approximation,

$$\bar{I} \sim e^{-(\eta + \eta_0)t} \left[\frac{\bar{\gamma}\ e^{\bar{\eta} t} + \bar{\gamma}_0\ e^{-\bar{\eta} t} - \Delta\ e^{-\Delta t}}{4\bar{\gamma}\bar{\gamma}_0\ (y^2 + \bar{\eta}^2)} + \frac{\Delta(e^{-\Delta t} - \cos yt) - y \sin yt}{(y^2 + \bar{\eta}^2)\ (y^2 + \Delta^2)}\right] , \qquad (2.14)$$

where $\Delta = 2\gamma_0 - \eta - \eta_0$. We can see that the shape of the line under consideration is determined by the parameters $\gamma_0 - \eta$ and $\eta - \eta_0$. From here it follows that in time-dependent spectra the compensation effect can be observed also for inhomogeneously-broadened ZPL. For example, if $|\eta - \eta_0| = \varepsilon$, $|\gamma_0 - \eta| = |\gamma_0 - \eta_0| = \varepsilon/2$, then a RSE line has the following shape:

$$\bar{I} \sim \frac{e^{-(\eta + \eta_0)t}}{\varepsilon(y^2 + \varepsilon^2)} \left(e^{\varepsilon t} - e^{-\varepsilon t} - 2\varepsilon y^{-1} \sin yt\right) . \qquad (2.15)$$

In the limit case of $\varepsilon \to 0$, one has $\bar{I} \sim y^{-2}\ e^{-2\gamma_0 t}\ (t - y^{-1} \sin yt)$, i.e. the half-width $\sim t^{-1}$.

A time-dependent absorption spectrum is determined as the dependence of the counting rate of scattered (emitted) photons on the excitation frequency Ω_0. It is easy to show that the formula for the absorption spectrum coincides with the one for the luminescence spectrum if one substitutes Ω and η by Ω_0 and η_0, respectively. Thus, in a time-dependent absorption spectrum, the compensation effect takes place as well, whereby, by selecting η_0 γ_0 and η_0 can be compensated.

In a number of cases, for example in the case of the Mössbauer effect, a resonant absorber (scatterer) is used instead of the spectrometer. Then in the limit case of a thin absorber [21]

$$C(\omega) = 1 - iC_0(\Omega - \omega + i\eta)^{-1} \quad , \tag{2.16}$$

where $C_0(\Omega - \omega + i\eta)^{-1}$ is the resonant part of a complex dielectric constant. If the radiation described by formula (2.7) is passed through such filter, then the time-dependent spectrum of the passed-through radiation is the following:

$$I(t) \sim e^{-2\gamma_0 t} - \frac{2C_0\, e^{-(\gamma_0 + \eta)t}}{(\Omega - \omega_0)^2 + (\gamma_0 - \eta)^2} \{(\Omega - \omega_0)\sin(\Omega - \omega_0)t$$

$$+ (\gamma_0 - \eta)\,[\cos(\Omega - \omega_0)t - e^{-(\gamma_0 - \eta)t}]\} \quad . \tag{2.17}$$

In this spectrum, the effect of the compensation of γ_0 and η takes place as well: if $\gamma_0 = \eta$, then $I \sim 1 - 2C_0(\Omega - \omega_0)^{-1}\sin(\Omega - \omega_0)t$. This formula, coinciding with the analogous formulae of [21] for the limit case of a thin absorber, describes a minimum with the halfwidth $\sim t^{-1}$, equal to the smallest value following from the uncertainty principle. In case of $\gamma_0 \sim \eta$ the transmission line is of an oscillatory form. Such type of a line is experimentally observable in the Mössbauer spectrochronograms [22].

The time-dependent spectrum considered above is measurable by the following scheme: the radiation is first passed through the spectrometer and then recorded at some time moment (more exactly, in a small interval from t to $t + \delta t$). In this way, a spectral analysis (filtration) of the radiation is carried out first and then the time is measured. By this scheme the measurement result depends essentially on the spectrometer resolution η that acts on the temporal qualities of the passed-through radiation.

However, another scheme is possible for measuring time-dependent spectra, where first the time is fixed (more exactly, the time interval from t to $t + \Delta t$) and only then the spectrum is measured [25]. To realize this scheme the time gate must evidently be used in front of the spectrometer (not behind it as in the previous scheme). Here the spectrometer operates like in the stationary case. Therefore, small η are the best for getting narrow lines. A time-dependent spectrum of such type is described by formulae (2.2) and (2.3) if by $C(t)$ the gate transmission function is meant. It is easy to show [25] that in this scheme the reverse time of the gate transmission $(\Delta t)^{-1}$ plays the same role as spectral resolution does in the first scheme. In particular, here, too, a compensation effect takes place; by a proper selection of Δt the broadening $(\Delta t)^{-1}$ and γ_0 can be compensated. Above the exponential transmission function C(t) (Lorentzian filtration function) was regarded. However, it is possible to show that, though incomplete, the compensation effect takes place for other transmission functions as well.

The compensation effect is manifested also in the narrowing of a hole in the inhomogeneous distribution function of the frequency of the optical transition between the ground and first excited level under a two-step pulse hole burning in a three-level system [26,27]. The total compensation of the natural width γ_0 of the first excited state and of the spectral width η_0 of the first pulse is possible in case of an exponentially decaying coherent first pulse (the Lorentzian function of frequency distribution) and the second limitingly short pulse. In this case, the shape of the hole mentioned is determined by formula (2.8), where t is the time interval between the maxima of the pulses.

References

1 K. K. Rebane: Elementarnaya teorija kolebatel'noj struktury spektrov primesnykh tsentrov kristallov (Nauka, Moskva 1968) [English transl.: Impurity Spectra of Solids (Plenum Press, New York 1970)]

2 M. Lax: J. Chem. Phys. 20, 1752 (1952)

3 R. Kubo, Y. Toyozawa: Progr. Theor. Phys. 13, 160 (1955)

4 I.S. Osad'ko: Usp. Fiz. Nauk 128, 31 (1979) [English transl.: Sov. Phys. - Usp. 22, 311 (1979)]; I.S. Osad'ko: in Spectroscopy and Excitation Dynamics of Condensed Molecular Systems, ed. by V.M. Agranovich, R.M. Hochstrasser (North-Holland, Amsterdam 1983) p. 437

5 J.L. Skinner, D. Hsu: J. Phys. Chem. 90, 4931 (1986).

6 G.F. Levenson: Phys. Status Solidi b43, 739 (1971)

7 V. Hizhnyakov: Phys. Status solidi b114, 721 (1982)

8 V. Hizhnyakov: in Proceedings of the Tenth International Conference on Raman Spectroscopy, ed. by W.L. Peticolas, B. Hudson (University of Oregon, Eugene 1986) p. 17-1; V. Hizhnyakov, I. Tehver: J. Raman Spectrosc. (in print)

9 V. Hizhnyakov, I. Tehver: Opt. Commun. 32, 419 (1980)

10 D.E. McCumber: J. Math. Phys. 5, 508 (1964)

11 M.A. Krivoglaz: Fiz. Tverd. Tela 6, 1707 (1964) [English transl.: Sov. Phys. - Solid State 6, 1340 (1964)]

12 V.Hizhnyakov, I. Tehver: Phys. Status Solidi b21, 755 (1967)

13 R.H. Silsbee: Phys. Rev. 128, 1726 (1962)

14 V.V. Hizhnyakov: in Teoriya lokal'nykh tsentrov kristalla, ed. by I.J. Tehver, Trudy IFA AN ESSR, Vol. 29 (Estonian SSR Acad. Sci., Tartu 1964) p. 83

15 V.V. Khizhnyakov, I.K. Rebane: Zh. Eksp. Teor. Fiz. 74, 885 (1978) [English transl.: Sov. Phys. - JETP 47, 463 (1978)]

16 I.K. Rebane, A.L. Tuul, V.V. Khizhnyakov: Zh. Eksp. Teor. Fiz. 77, 1302 (1979) [English transl.: Sov. Phys. - JETP 50, 655 (1979)]

17 F. Shimizu, K. Umezu, H. Takuma: Phys. Rev. Lett. 47, 825 (1981); F. Shimizu, K. Shimizu, H. Takuma: Phys. Rev. A28, 2248 (1983)

18 W. Zinth, M.C. Nuss, W. Kaiser: Opt. Commun. 44, 262 (1983)

19 P.E. Coleman, D. Kagan, P.L. Knight: Opt. Commun. 36, 127 (1981)

20 J.H. Eberly, C.V. Kunasz, K. Wodkiewicz: J. Phys. B13, 217 (1980)

21 F.J. Lynch, R.E. Holland, M. Hamermesh: Phys. Rev. 120, 513 (1960)

22 R. Koch, E. Realo: Eesti NSV Tead. Akad. Toim. Füüs. Matem. (Proc. Estonian SSR Acad. Sci.) 28, 374 (1979); 30, 171 (1981) (in English)

23 I.-J. Ma, J. Mertens, G. zu Putlitz, G. Schütte: Z. Phys. 208, 352 (1968)
24 A.B. Doktorov, O.A. Anisimov, A.I. Burshtein, Yu.N. Molin: Chem. Phys. 71, 1 (1982)
25 I. Rebane, V. Hizhnyakov: Eesti NSV Tead. Akad. Toim. Füüs. Matem. (Proc. Estonian SSR Acad. Sci.) 30, 1 (1981) (in Russian)
26 I. Rebane: Eesti NSV Tead. Akad. Toim. Füüs. Matem. (Proc. Estonian SSR Acad. Sci.) 35, 296 (1986); 35, 400 (1986) (in Russian)
27 I. Rebane: Eesti NSV Tead. Akad. Toim. Füüs. Matem. (Proc. Estonian SSR Acad. Sci.) 36, 204 (1987) (in Russian)

Zero-Phonon Lines in the Spectra of Small Impurity Molecules as an Indicator of Electron-Phonon Interactions in Crystals

A. Freiberg and L.A. Rebane

Institute of Physics, Estonian SSR Acad. Sci.
SU-202400 Tartu, Estonian SSR, USSR

Abstract

Low-temperature spectra of small molecular anions O_2^-, S_2^-, NO_2^- in alkali halide crystals are reviewed as model systems for the theory of the spectra and electron-phonon interactions of impurity centres in crystals. Zero-phonon lines of intramolecular transitions as well as the dependences of their contours on the crystal's temperature and defect concentration have been analyzed. The participation of the impurity's rotational and librational modes in the relaxation of the excited state and in the broadening of zero-phonon lines has been elucidated.

1. Small Molecular Centres as Model Systems for the Theory of the Spectra of Impurity Centres in Crystals

The study of the optical spectra of two- and three-atom molecular anions (O_2^-, S_2^-, NO_2^- and others) embedded in alkali halide crystals (AHC) was initiated at the Institute of Physics of the Estonian SSR Academy of Sciences in 1964 by a group of three: A.Laisaar, L.A.Rebane and T.Saar. The idea was proposed by K.Rebane with the aim of comparing the observed vibronic spectral structure with the one expected theoretically on the basis of the theory of impurity centres in crystals when the centre displays a high-frequency local vibration [1]. In case of a strong interaction of the electronic transition with the high-frequency local vibration in the centre and a relatively weak interaction with crystal phonons, the theory predicts a characteristic vibrational structure of the spectrum consisting of a purely electronic line (PEL) and a number of zero-phonon lines (ZPL) which are replicas of PEL on high-frequency local vibration. Each narrow ZPL has to be adjoined by a broad phonon wing that is due to transitions in which, along with local impurity excitations, also non-localized vibrations of the crystal-matrix, phonons, are involved. It was expected that in case of molecular impurities the role of high-frequency local vibrations should be performed by the intramolecular vibrations. For light molecules their frequencies essentially exceed the limiting phonon frequencies and, therefore, no overlapping of the phonon wing with the subsequent ZPL should occur.

The O_2^- in KCl whose luminescence spectrum, observed earlier by ROLFE et al. [2], revealed a complicated structure, was proposed by A.Laisaar as a suitable object. With the help of T.Saar polycrystallic samples of different AHC with O_2^- centres were prepared (later, T.Avarmaa organized the growing of high-quality alkali halide monocrystals doped with a variety of other molecular anions). The luminescence spectra of O_2^- in AHC at 4.2 K (see Figure 1) were measured by L.A.Rebane at first in the

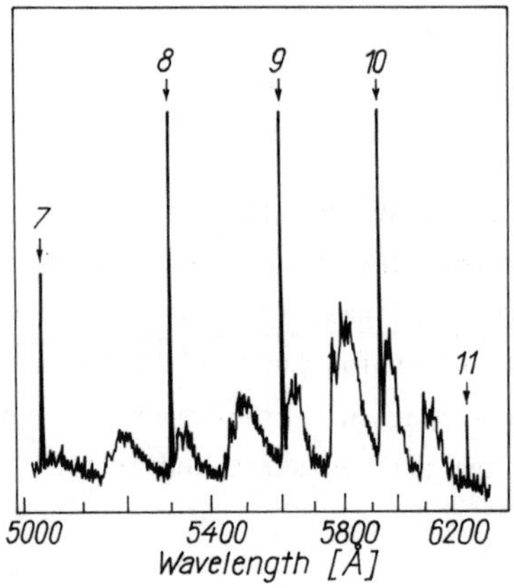

Figure 1: KCl:O_2^- luminescence spectrum at 4.2 K [3]. The quantum numbers of the intramolecular vibration of the ground electronic state are indicated

laboratory of E.F.Gross in the Leningrad Physico-Technical Institute in collaboration with B.P.Zakharchenya. The spectra really revealed series of quite narrow explicit lines and broad structural bands that were interpreted in [3], by using the theory of [1], as ZPL and phonon wings in the series of intramolecular electron-vibrational transitions. The intramolecular vibration of O_2^- performs a role of a high-frequency local vibration of the centre. Paper [3] presented the first experimental proof of the validity of the theory of the impurity centre with a high-frequency local vibration. It was not until 1970 that the phonon wings in the spectra of the Shpol'skii systems - certain organic impurity molecules in molecular cyrstals - were found [4], which demonstrated the validity of the theoretical description of the spectra of molecular centres regardless of their chemical nature.

In subsequent papers, the spectra of O_2^-, S_2^-, Se_2^-, NO_2^-, PO_2^-, SH^- and other centres in various AHC were studied. The characteristic structure containing a set of ZPL and phonon wings was found and thoroughly studied in the spectra of luminescence, absorption (see reviews [5-7]) as well as in the spectra of infrared absorption [8,9], Raman scattering [10,11], and hot luminescence [12].

In Figure 2, a complicated vibronic structure of the absorption and luminescence bands, which are due to the $^1B_1 - {}^1A_1$ electronic transition in the impurity molecule of NO_2^- in CsI crystal, is depicted [13]. The observed ZPL correspond to vibronic transitions where two totally symmetric molecular vibrations of angular frequencies ω_1 and ω_2 are excited (the quantum numbers of the final vibronic state, v_1 and v_2,

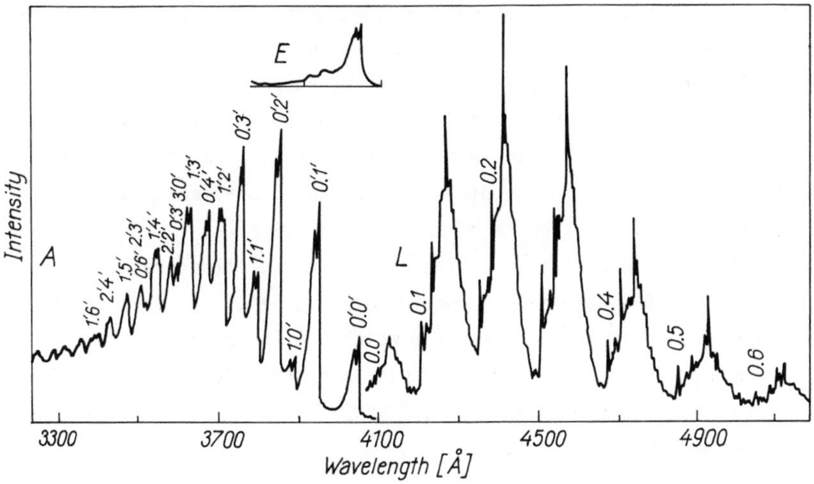

Figure 2: The luminescence (L), absorption (A), and luminescence excitation spectra (E) corresponding to the electronic transition $^1B_1 \leftrightarrow {}^1A_1$ in NO_2^- centre in CsI crystal at 4.2 K [13]

are indicated in Figure 2 at the corresponding ZPL).

The following aspects of the molecular impurity centres ensured their success as model systems for the theory of impurity spectra of solids [14]:

1. The molecular anions listed above form stable luminescent centres in AHC. The vibrational as well as the lower excited electronic states of an impurity molecule appear to be well separated from the host crystal states. It gives rise to impurity spectra which largely retain the well-known spectra of free molecules, thus essentially simplifying the interpretation.
2. Relatively high frequencies of molecular vibrations and rather strong electron-vibrational coupling, which is characteristic of light molecules, lead to the appearance of extended series of well-resolved vibronic bands in the spectra of impurity molecules. This allows one (i) to study the "similarity law" of the basic theory [1], according to which the spectral distribution has to be the same for each vibronic band, and (ii) to refine the theoretical model of the centre.
3. High intensity of ZPL and their good separation from the accompanying broad sidebands is another important feature of the spectra of molecular centres. In many cases, the intensity distribution in the sideband reproduces well the function of the phonon density of states of the host crystal. It makes obvious the interpretation of the sideband as phonon-assisted transitions (phonon wings). The Debye-Waller factors usually observed for the molecular centres are of the order of 0.1.
4. Owing to the high quality of the impurity crystals under study, the inhomogeneous broadening of the spectra of molecular centres is relatively small (0.5 -

5 cm^{-1}) and it does not smear out the characteristic vibronic structure of the homogeneous spectrum. Note that the inhomogeneous spread of the electronic transition energies of impurity organic molecules in organic crystals is typically about 100 cm^{-1}, that exceeds the homogeneous width of a PEL by $\sim 10^6$ times. It is the huge inhomogeneous broadening that has for a long time been the main obstacle in the way of detecting the theoretically-predicted vibronic structure in impurity spectra of organic molecules [15].

A clear physical interpretation of the main vibronic structure of the spectra of molecular centres in AHC has enabled one to obtain a detailed spectroscopic information about the very molecular-radicals for which AHC appeared to be a good stabilizing matrix. Molecular anions of a more complex structure (e.g. SO_2^{2-}, $S_2O_3^{2-}$, etc.) are easily stabilized in frozen water solutions of alkali halide salts and simple acids, forming luminescent centres [16,17].

2. Some Aspects of the Spectroscopy of Small Molecular Centres

A detailed spectroscopic study of small molecular anions in AHC has resulted in a more comprehensive understanding of electron-vibrational interactions in molecular impurity centres, which in turn has assisted the development of new branches in the theory.

The study of the luminescence yield in dependence on temperature and excitation frequency has explicitly revealed the high-frequency local vibrations play an important role in the relaxation of excitation energy to phonons [18-20]. The step-by-step decrease of the luminescence yield on exciting the centre to higher vibronic levels of the vibration ω_2 in $^1B_1$ electronic state of NO_2^- impurity in KCl crystal is depicted in Figure 3. In this case two processes of the excited state radiationless deactivation - the anharmonic decay of the local vibrational excitation, ω_2, and the electronic cross-relaxation - are of comparable rates both, however, being much slower than the relaxation of the phonon contribution of excitation. This leads to an unusual stepwise dependence of the luminescence yield on the excitation frequency.

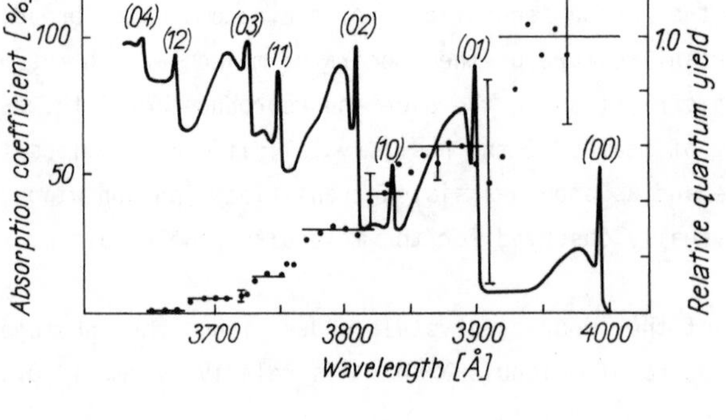

Figure 3: The absorption spectrum (continuous line) and relative quantum yield of the luminescence (dots) of NO_2^- centres in KCl at 4.2 K [20]

A weak emission conditioned by the nonequilibrium population of vibronic levels, the so-called hot luminescence (HL), has been observed in case of NO_2^- in KCl [21]. In Figure 4, the dependence of the HL spectrum on the excitation frequency is shown to characterize the variations of the nonequilibrium populations of the vibronic levels of NO_2^- molecule in dependence on which of the above-lying vibronic levels is populated by direct excitation. The recording of HL line intensities under stationary laser excitation enabled one to plot a detailed picture of vibrational relaxation in

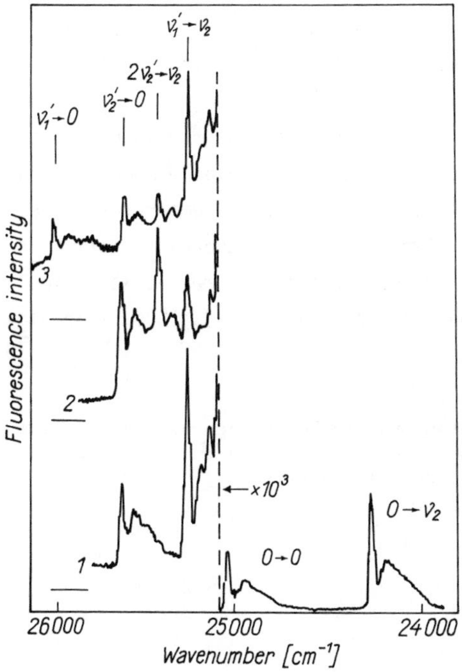

Figure 4: The hot luminescence spectrum of NO_2^- in KCl in the anti-Stokes region of vibrationally equilibrium (ordinary) luminescence spectrum (OL) at various excitation light wavelengths [12]. The corresponding vibronic transitions are indicated above the lines

the excited electronic state and derive the picosecond relaxation times of particular levels. The HL method, along with the direct time-resolved measurements under the excitation of a centre by pulses of picosecond duration [22,23], is now widely used in the studies of fast relaxations in condensed matter (see reviews [12,24]). These experiments have stimulated the development of the theory of the resonant secondary emission of the impurity centre.

Molecular centres of low symmetry occupy a number of equivalent positions in the AHC lattice, the reorientations between which may be expressed as rotational degrees of freedom. This kind of pseudorotational motion has been studied by the absorption, luminescence and Raman scattering spectra [25-29]. For example, NO_2^- centres in KCl, KBr and RbCl crystals perform a slightly hindered one-dimensional rotation around the molecular axis of the minimum inertia moment. This motion manifests itself in a peculiar rotational structure of ZPL [25] and in the decrease of the luminescence polarization degree. In case of strong static coupling with the surrounding lattice, like in case of NO_2^- in KI, the rotation is frozen and the molecular centre performs librations around the fixed axes of the lattice symmetry, accompanied by occasional reorientations of molecular axes. This kind of motion is revealed in vibronic spectra by librational lines [30] and it causes changes in the polarization of luminescence [31-34] as well as in the Raman scattering spectra [28]. By recording the degree of luminescence polarization in dependence on time [35,36], excitation frequency [31-33] and temperature [31,32,34,37], the elementary mechanisms of reorientation of molecules during the lifetime of the excited electronic state have been distinguished (see reviews [7,38]). The reorientation of molecules in the ground electronic state has been studied by following the decay kinetics of the induced linear dichroism [39].

In Figure 5, the temperature dependence of the polarization degree is presented for O_2^- in KCl and RbBr. The reorientation probabilities of the dipole moment of the electronic transition (hence, that of the molecular axis) versus temperature have been determined from these data. As a result, two processes of thermally activated reorientations can be distinguished, from which the low-temperature reorientation takes place when the librational level is thermally populated [33]. Thus, librational motion is in close interaction with phonons.

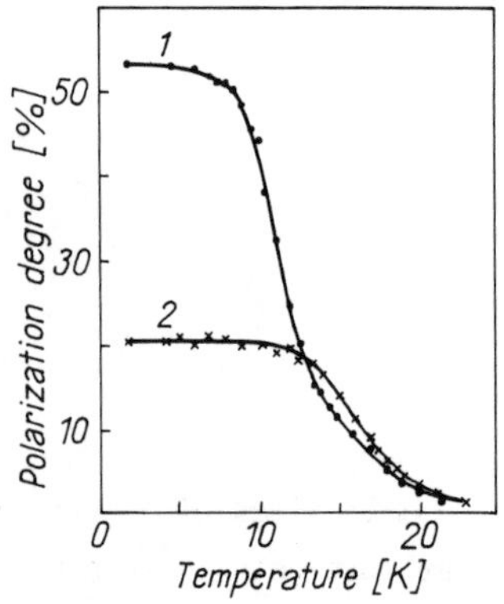

Figure 5: Temperature dependence of the polarization degree of O_2^- luminescence in KCl (curve 1) and RbBr (curve 2) crystals [33]

The participation of librations in the energy exchange between the centre and the crystal at low temperatures causes anomalies also in the temperature broadening of ZPL in the spectra of molecular centres. These problems will be discussed in more detail below.

The impurity molecule affects strongly the vibrations of the surrounding crystal atoms. A comprehensive understanding of the local vibrational dynamics has been achieved by studying the ZPL phonon sidebands in the absorption and luminescence spectra of a number of molecular centres [5,13,19,25,40,41] as well as the impurity-induced one-phonon Raman scattering spectra of AHC [11,42].

In Figure 6, a phonon sideband in the luminescence spectrum of O_2^- in KI is shown. In this case, the structure of the ZPL sideband corresponds well to the density of one-phonon states of a pure KI crystal (shown in the upper part of the Figure), indicating a prominent contribution of the phonons of F_{2g} symmetry. Low-frequency

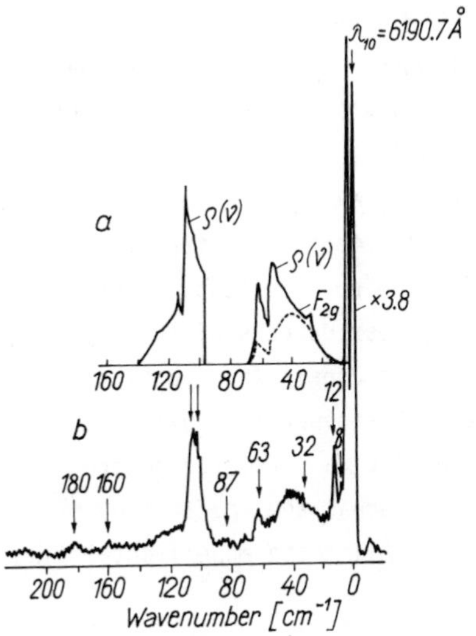

Figure 6: The phonon sideband of the $v = 10$ ZPL in the luminescence spectrum of O_2^- centres in KI crystals at 4.2 K [40]. $\rho(\nu)$ is the phonon spectrum of an ideal KI crystal

pseudolocal modes, arising from molecular librations, as well as a gap mode are also displayed. In such cases, when the local dynamics of the crystal is only slightly affected by the impurity centre, the structure of phonon sidebands may serve as a good source of information about the phonon spectrum of the matrix crystal. However, the low symmetry of molecular centres usually brings about substantial anisotropic changes in the crystal force constants and, subsequently, essential changes in the local dynamics of the crystal [42].

3. Inhomogeneous Broadening of ZPL and Evaluation of Homogeneous Spectrum

The spectra of small molecular centres in AHC recorded under nonselective excitation undergo an essential inhomogeneous broadening, which at low temperatures determines the contours of ZPL. The inhomogeneous distribution of the electronic transition energy of the centres reflects the distribution of site energies due to the local strain fields in the crystal [43]. It has incorrectly been assumed for a long time that all kinds of strain sources lead to Gaussian inhomogeneously-broadened ZPL contours. Stoneham has shown that, in general, it is not true [43]. Thus, the inhomogeneous distribution function (IDF) will provide important information about the sources of strains in the crystal that cause the ZPL broadening of probe impurity molecules.

It is interesting to note that small impurity molecules may occupy more than one crystal site. This provides an additional structure of the spectrum, which is like the multiplet structure in the Shpol'skii spectra. Such structure has been observed in the spectra of NO_2^- in CsCl [13]. In this case the IDF description has to be applied to each site component separately.

For molecular centres in AHC the point defects and linear dislocations turned out to be the main sources of inhomogeneity [44,45]. Figure 7 depicts the broadening and shift of ZPL in the luminescence spectrum of O_2^- in KCl at 4.2 K when the point defect concentration increases. The point defects are actually O_2^- centres themselves and the accompanying OH^- centres. The ZPL, having an approximate Voigt contour [46] at low concentrations, becomes Lorentzian when the point defect concentration is high, whereby the Lorentzian part increases linearly with the point defect concentration. The Gaussian part of the ZPL contour is related to linear defects of the crystal. This has been experimentally proved by creating various densities of dislocations in KCl [44,45]. The Gaussian width (here and throughout this paper the full width of the line at half intensity maximum is considered) of the ZPL was observed to grow linearly with the square root of the dislocation density, whereas the spectral position of the inhomogeneously-broadened ZPL remained unchanged. All these experimental findings are in qualitative accordance with the theory [43].

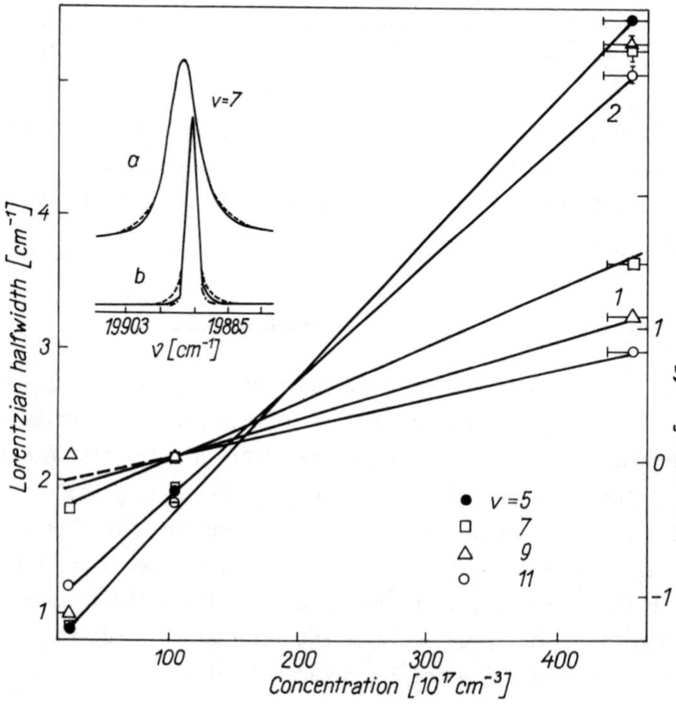

Figure 7: The shift of the maximum $\Delta\omega$ (1) and the Lorentzian width Γ_L (2) of ZPL with various quantum numbers v in $KCl:O_2^-$ luminescence spectrum at 4.2 K versus total concentration of point defects. In the insert, an approximation of the v = 7 ZPL contour in case of various defect concentrations by a Lorentzian (dashed line) and a Gaussian (dash-dotted line) curve [47]

The inhomogeneous effect on the energy of vibronic transitions turned out to depend on the level quantum number v of the excited vibrational state in the ground electronic state. This effect is clearly visible in Figure 7 where the point-defect-induced broadening and shift of ZPL are different for the transitions with $v = 5$, 7, 9, and 11. At a higher defect concentration the broadening and shift of ZPL with e.g. $v = 7$ are larger than those for ZPL with $v = 11$, whereas at low concentration the dependence is inverse. This interesting effect has no theoretical explanation yet. But qualitatively the situation can be understood by assuming that the local strain affects not only the electronic transition energy, but also the frequency of intramolecular vibration [47].

When the homogeneous and inhomogeneous contributions are comparable, the width of ZPL displays rather a complicated dependence on the vibrational number v. An example concerning the ZPL widths in the luminescence spectrum of O_2^- in CsBr is shown in Figure 8 [45,48]. The contribution of the homogeneous width was increased by rising the crystal temperature up to 15 K.

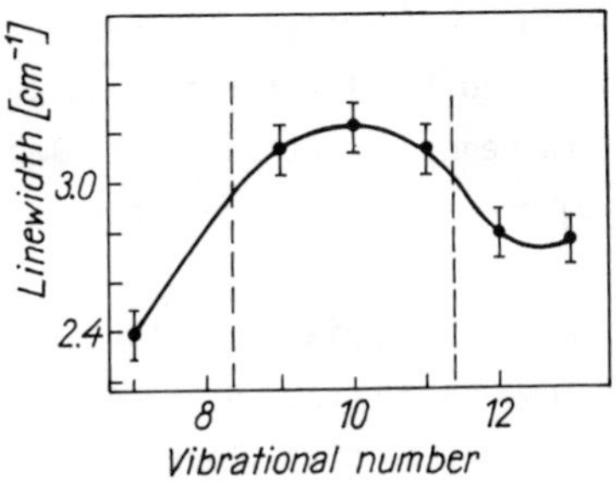

Figure 8: The width of a ZPL in CsBr:O_2^- spectrum at 15 K versus the vibrational quantum number v [45]

The inhomogeneous linewidths were decreased by annealing the crystals and by an extreme lowering of the impurity concentration. It makes possible to distinguish the homogeneous vibronic ZPL widths of O_2^- centre at 4.2 K to be as small as 0.1 cm^{-1}, which, in the case of a lifetime-limited broadening, corresponds to about 50 ps anharmonic decay time of the intramolecular local vibration in the ground electronic state. These rather narrow linewidths make vibronic ZPL, along with PEL, sensitive probes for studying the crystal structure and the interactions of optical electrons of the centre with localized and delocalized vibrations of the crystal lattice.

4. Homogeneous Contour and Temperature Broadening of ZPL

The homogeneous contour of ZPL reflects the dynamic influence of the lattice on the probe molecule. In the case of model two-level electronic transition and adiabatic approximation the difference between vibrational Hamiltonians in the ground and excited states is responsible for the broadening of ZPL and its expansion to the second and higher orders with respect to phonon operators in a harmonic crystal should be considered [49-51]. The studying of homogeneous contours and their dependence on temperature in the region much less than the Debye temperature T_D of the matrix provide experimental data for elucidating electron-phonon broadening mechanisms. In the last decade a number of experimental studies have been made (see, e.g. reviews [52-57]) on the PEL and vibronic ZPL in the spectra of ions and molecules in crystals as well as in polymers and glasses by using both the traditional spectroscopic methods and the more advanced methods of selective laser spectroscopy (spectral hole burning and fluorescence line narrowing). This, in its turn, stimulates the further development of the ZPL theory considering also the coupling of impurity centre transitions with low-frequency matrix vibrations [53,56,58-61]. The main trigger of the high theoretical activity in the last few years has been the anomalous behaviour of the homogeneous broadening of impurity ZPL in amorphous hosts. Here low-temperature linewidths are by one to three orders of magnitude broader than for the same impurities in crystals. Also, temperature dependences are much weaker, being linear to quadratic functions of temperature rather than exponential or T^7 [56,57].

It is interesting to recall here the results of the vibronic ZPL broadening in the spectra of small molecular centres, in which the main peculiarities of the electron-phonon coupling of the impurity molecules were observed for the first time. In particular, the participation of pseudolocalized low-frequency vibrations (presumably the librations of an impurity molecule) in dephasing processes has been shown [30, 62,63].

The ZPL contours in the luminescence spectra of O_2^- and S_2^- impurities in two alkali halide crystals at 4.2 K are presented in Figure 9. Low concentration of controlled impurities ($\leq 3\cdot 10^{18}$ cm^{-3}) together with a negligible density of dislocations lead to a nearly-Lorentzian contour with a small asymmetry (the intensity fall is less steep towards lower frequencies). The asymmetry seems to become smaller when the defect concentration is increased.

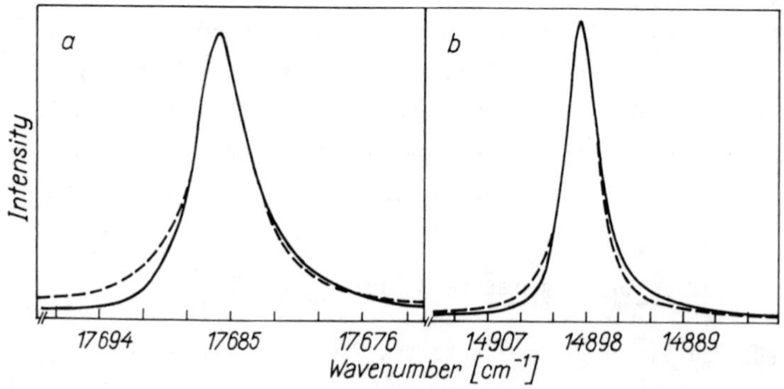

Figure 9: The ZPL contours of the luminescence spectra of O_2^- and S_2^- centres. The approximation (dashed line) has been performed with the help of a symmetric convolution of a Lorentzian profile with a Gaussian profile ($\Gamma_L/\Gamma_G = 6.6$ and 2.5 in case of KI:S_2^- and CsBr:O_2^-, respectively) [47]

It is worth stressing that this important aspect of the ZPL shape that, as we guess, reflects very fine details of the impurity-lattice interactions, has been poorly discussed in literature. Most of the popular models [50,53,58,59] ignore the problem totally. As has been shown by other authors, the possible reasons for the lineshape asymmetry may be quadratic electron-phonon coupling [61,64], anharmonicity of vibrations [51] as well as non-Markovian character of relaxation.

With the rise of temperature all ZPL display broadening, peak shift, $\Delta\omega$, and some increase in asymmetry (see Figure 10). The Lorentzian linewidth Γ and shift $\Delta\omega$ as a function of temperature for the vibronic transition $v' = 0 \rightarrow v = 7$ ZPL in the luminescence spectrum of O_2^- in KCl are depicted in Figure 11. The dependence of $\Gamma(T)$ is nearly quadratic in the whole temperature region, 4.2 - 50 K, which, at the first sight, is unexpected within the frames of the quadratic vibronic coupling theory [50,51,53] and with the fact that in AHC $T_D \geq 100$ K taken into account.

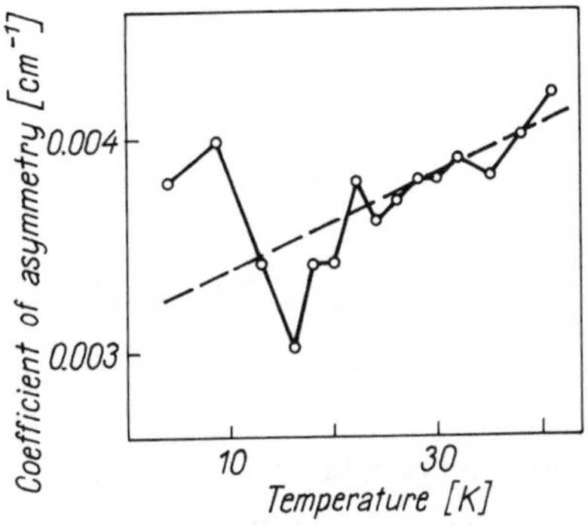

Figure 10: Temperature dependence of the coefficient of asymmetry of $v = 7$ ZPL in the luminescence spectrum of $KCl{:}O_2^-$ crystal [47]

The homogeneous width of a ZPL is determined by the relaxation time T_2 of the excited electronic state of the impurity in a matrix and can be written: $\Gamma = (\pi T_2)^{-1} = (2\pi T_1)^{-1} + (\pi T_2')^{-1}$, where T_1 is the population relaxation time; T_2', the "pure" dephasing time, is determined by fluctuations in the optical transition frequency.

The anharmonic decay of the final vibrational state along with the decay of the initial electronic state are responsible for the residual homogeneous ZPL widths in our spectra within the limit $T = 0$. In practice, an inhomogeneous contribution should also be considered. In Figure 11, e.g., $\Gamma(0) \approx 2.5$ cm^{-1}. The contributions of radiational and radiationless decay of the initial electronic state to this residual width is negligible ($0.5 \cdot 10^{-4}$ cm^{-1} as estimated from the luminescence decay time 10^{-7} s). As the frequencies of the ground state molecular vibrations of small molecules exceed for many times maximum phonon frequencies in AHC (5 times for O_2^- in KCl), the probability of the anharmonic decay of local vibrational excitation to phonons is small and the process makes no essential contribution to the observed dependence $\Gamma(T)$ at $T < T_D$. This conclusion was recently confirmed by the measurements of the temperature dependence of CN^--impurity vibrational fluorescence lifetime in a number of alkali halide hosts [65]. Finally, as the initial electronic state lifetime is also almost constant in the actual temperature region, we come to a conclusion that the only possibility is that the observed dependence $\Gamma(T)$ is mostly due to the electronic excitation dephasing, which is described by the quadratic electron-phonon cou-

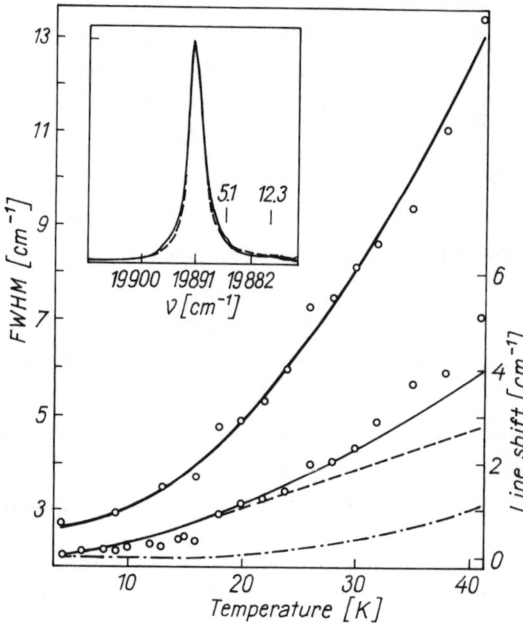

Figure 11: Temperature dependence of the width Γ and the shift Δω of the v = 7 ZPL contour in the O_2^- luminescence spectrum in KCl. The continuous line is ZPL width approximated by formula (1). 1 - anharmonic shift of ZPL due to the thermal broadening of the crystal; 2 - harmonic shift of ZPL due to the quadratic electron-phonon interaction; 3 - total calculated lineshift Δω . In the insert, the v = 7 ZPL contour is shown. Approximation is performed by an asymmetric Lorentzian curve (dashed line) [63]

pling terms of the interaction operator. The dephasing (or modulation, as it is called in [51]), may be considered as the result of quasi-elastic phonon scattering by impurities. In case of the Debye phonon distribution it leads to the dependence $\Gamma \sim T^7$ at $T < T_D/2$ and to $\Gamma \sim T^2$ at high temperatures ($T \geq T_D$).

To combine the theoretical predictions with the observed quadratic dependence of ZPL widths at low temperatures starting from 4.2 K we interpreted the latter in [30] as a result of modulational broadening of impurity levels due to interaction with a low-frequency pseudolocal vibration. In accordance with this, the ZPL broadening has been calculated by a simple formula [51]

$$\Gamma(T) = \Gamma(0) + a\, n(\omega_o)[n(\omega_o) + 1] \,, \tag{1}$$

where $n(\omega_o)$ is the number of pseudolocal phonons at thermal equilibrium, ω_o is their frequency and $\Gamma(0)$ is the residual linewidth. The pseudolocal vibrations with the energies of about 10 cm^{-1} were found in the luminescence spectra of O_2^- in KCl, KBr and KI as well as for S_2^- in KI. The calculation of $\Gamma(T)$ for O_2^- in KCl according to formula (1) with $\hbar\omega_o = 12.3$ cm^{-1} is presented in Figure 11. Similar calculations of the temperature broadening were performed for other systems studied which gave rather a good agreement with the experimental data [62,63]. As the ZPL asymmetry is nearly proportional to $\Gamma(T)/T$ [61,64], there is a qualitative consistence also with the asymmetry data (see Figure 10).

The idea of low-frequency pseudolocal vibrations (librations) carrying the better part of quadratic electron-phonon interaction and impurity electronic excitation dephasing at low temperatures, originally proposed in [30], is now widely accepted, which allows the interpretation of the experimental data on the temperature broadening of narrow spectral ZPL holes and on the temperature decay of photon echo signal in crystals (see, e.g. reviews [54,58]). The dephasing by the very low frequency vibrations localized not directly at the impurity site, presumably actual in glassy hosts, has also been taken into account in a number of recent theoretical models (see, e.g. [56,60]).

5. Integral Intensity of ZPL

The inhomogeneous broadening of ZPL, which does not wash out separation between ZPL and its phonon sideband, is no obstacle to the experimental study of such an important spectral characteristic as the Debye-Waller factor (DWF). Within the frames of the basic model of impurity centres in crystals [1] the DWF, α, is properly connected with the parameters of linear electron-phonon coupling - the shifts of the equilibrium coordinates, q_{oi}, of the vibrational oscillators on electronic excitation in the centre:

$$\alpha(T) = I_o/(I_o + I_p) = \exp[-S(T)] = \exp[-\sum_{i=1}^{N}(n_i + 1/2)q_{oi}^2] . \qquad (2)$$

Here I_o and I_p are the integral intensities of the measured ZPL and the corresponding phonon wing; n_i is the temperature-dependent number of phonons of frequency ω_i; S(T) is the so-called total dimensionless Stokes losses on crystal vibrations.

S(T) and its dependence on temperature, impurity concentration and on the number v of vibronic group was studied in the luminescence spectra of impurity molecules of O_2^- in KCl, KI and CsBr, S_2^- in KI and NO_2^- in KCl [66,67]. The experimental results obtained for O_2^- in KCl are presented in Figure 12 together with the calculated curves of S(T). To explain the observed dependence S(T) it was necessary to take into account not only the acoustic Debye phonons but also to add the contribution of the pseudolocal vibration that determined the corresponding ZPL broadening as shown before.

Figure 12 demonstrates another interesting effect - the decrease of the Stokes losses S on the vibronic group number v. The decrease of the total Stokes losses is accompanied also by the decrease of the multiphonon contribution to the phonon wings. Thus, there is definitely a reduction of the shifts of phonon oscillators when higher levels of the ground state molecular vibration are excited. The same effect has been observed at higher temperatures as the narrowing of vibronic bands [68]. It indicates an anharmonic interaction between intramolecular and crystal vibrations and thus the violation of the "similarity law" [1]. The S(v) dependence has been successfully interpreted with the help of the double adiabatic approximation method and with the third-order anharmonicity taken into account [66]. In this approximation, S(v) can be described as follows:

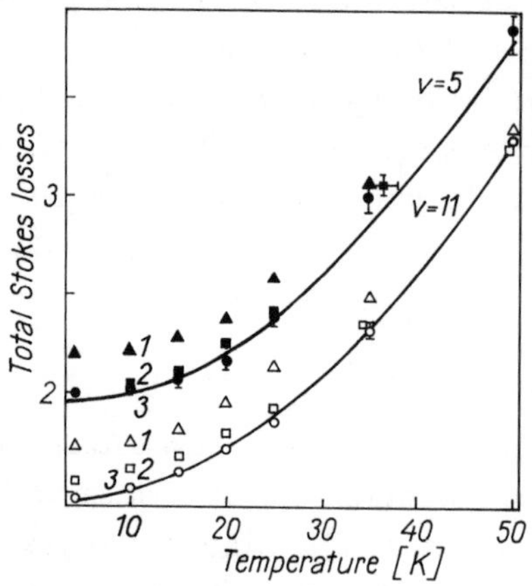

Figure 12: Temperature dependence of the total Stokes losses on crystal vibrations in the luminescence spectrum of KCl:O_2^- crystal in case of various levels of excitation of the intramolecular vibration v and at different O_2^- impurity concentrations. Concentration of O_2^- centres: $1.5\cdot10^{19}$ cm^{-3} (1); $1.5\cdot10^{18}$ cm^{-3} (2); $2.5\cdot10^{17}$ cm^{-3} (3). Solid curves - theoretical adjustment [47]

$$S(v) = (1/2) \sum_i [q_{0i} - (v + 1/2)d_i/\hbar\omega_i]^2 , \tag{3}$$

where d_i is the anharmonic coupling coefficient in the i-th vibrational mode. The S(v) calculations by formula (3), when phonons were replaced by an effective oscillator (see Figure 13, solid curves), are in good accordance with the experimental data, including the minimum near $v = 9$.

The increase of the total Stokes losses with the growth of the impurity concentration is observed in Figures 12 and 13 [47]. It indicates the change of the DWF along the inhomogeneous contour, i.e. the DWF is different for unequivalent centres. This fine effect has lately been observed for some organic impurities in organic matrices as well by using the selective laser spectroscopy methods [69]. The energy transfer

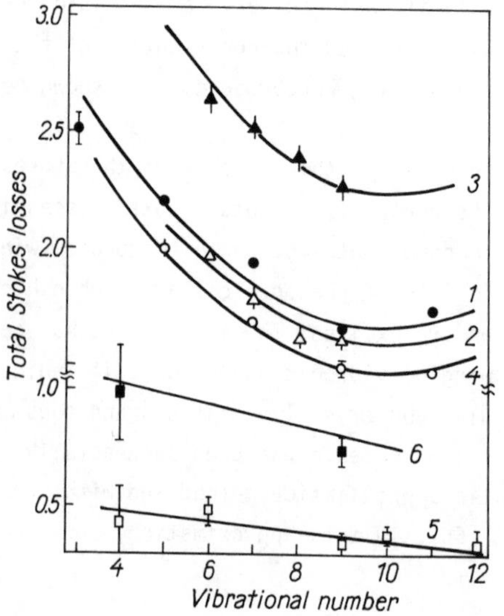

Figure 13: Total Stokes losses versus the number of the vibronic group v, temperature and impurity concentration. Solid curves - theoretical adjustment by formula (3). Curves 1 - 4 for KCl:O_2^-; 5, 6 for CsBr:O_2^-; 1 - T = 4.2 K, $1.5\cdot10^{18}$ cm^{-3}; 2 - T = 6, $8\cdot10^{18}$ cm^{-3}; 3 - T = 29 K, $8\cdot10^{18}$ cm^{-3}; 4 - T = 4.2 K, $2.5\cdot10^{17}$ cm^{-3}; 5 - T = 4.2 K, $1\cdot10^{18}$ cm^{-3}; 6 - T = 50 K, $1\cdot10^{18}$ cm^{-3} [67]

between inhomogeneous centres as the possible reason of the change of DWF with impurity concentration should also be considered [70,71].

6. Concluding Remarks

The problem of optical linewidths and the study of dephasing processes has received a renewed interest in recent years mainly due to the development of laser sources and the new laser-based methods of investigation in frequency as well as in time domain. It seems that further progress can be made when spectrochronography [22,23,61,72], that combines the frequency- and time-domain methods, is applied. Interesting new effects revealing the intimate aspects of light-matter interactions and excitation relaxation are to be expected [61,72].

Small molecular ions in AHC have proved to be relatively simple model impurity systems possessing a very rich spectrum of interesting and useful properties. Let us list here only some of them which have been under discussion during the last few years:

1. O_2^- in KCl is one of the first solid state systems in which the collective spontaneous emission of coherent light - superfluorescence - has been observed [73].
2. Intensive infrared vibrational emission of CN^- defects in a number of alkali halide hosts has been discovered [74] and extensively studied [65]. As a result, for the first time a laser generation based on vibrational transitions of a molecular defect isolated in a solid matrix has been obtained (for the transition $v = 2 \rightarrow v = 1$ of a CN^- impurity [75]).
3. On inhomogeneously-broadened vibrational transitions of ReO_4^- in AHC spectral hole burning has been demonstrated [76], which provides the opportunity to study the homogeneous vibrational lineshape.
4. An efficient and stable colour centre laser operation has been found in additively-coloured NaCl, KCl and KBr crystals doped with anionic molecular impurities. $NaCl{:}OH^-$ crystals [77] and $KCl{:}Na^+{:}O_2^-$ crystals [78] allow the tuning of the lasing wavelength over the wide range from 1.42 to 1.72 μm and from 1.71 to 2.21 μm, respectively.
5. The fine vibronic structure of the spectra of small molecular ions has turned out to be very useful in resolving the widely-discussed Raman scattering/fluorescence problem at resonant excitation [24,79] and in examining the effect of high hydrostatic pressure on the spectra of impurity molecules [80].

References

1 K.K. Rebane: Elementarnaya teoriya kolebatel'noj struktury spektrov primesnykh tsentrov kristallov (Nauka, Moskva 1968) [English transl.: Impurity Spectra of Solids (Plenum Press, New York 1970)]
2 J. Rolfe, F.R. Lipsett, W.J. King: Phys. Rev. 123, 447 (1961)
3 K. Rebane, L. Rebane: Eesti NSV Tead. Akad. Toim. Füüs. Matem. Tehn. (Proc. Estonian SSR Acad. Sci) 14, 309 (1965) (in Russian)
4 K. Rebane, P. Saari, T. Tamm: Eesti NSV Tead. Akad. Toim. Füüs. Matem. (Proc. Estonian SSR Acad. Sci.) 19, 251 (1970) (in Russian)
5 L.A. Rebane: in Physics of Impurity Centres in Crystals, ed. by G.S. Zavt (Estonian SSR Acad. Sci., Tallinn 1972) p. 353
6 K.K. Rebane, L.A. Rebane: Pure Appl. Chem. 37, 161 (1974)
7 L.A. Rebane, O.I. Sild: in Defects in Insulating Crystals, ed. by V.M. Tuchkevich, K.K. Shvarts (Zinatne, Riga, Springer, Berlin, Heidelberg, New York 1981) p. 619
8 T.H. Mauring: Ph.D. Thesis, Inst. Phys., Estonian SSR Acad. Sci., Tartu (1973)
9 A.I. Stekhanov, T.I. Maksimova: Fiz. Tverd. Tela 9, 2590 (1967)
10 T.P. Martin: Phys. Rev. B11, 875 (1975)
11 L. Rebane, K. Haller, T. Haldre, A. Novik: Eesti NSV Tead. Akad. Toim. Füüs. Matem. (Proc. Estonian SSR Acad. Sci.) 24, 107 (1975) (in Russian)
12 K. Rebane, P. Saari: J. Lumin. 16, 223 (1978)
13 A. Freiberg, P. Kukk: Chem. Phys. 40, 405 (1979)
14 L.A. Rebane: Dr. Thesis, Inst. Phys., Estonian SSR Acad. Sci., Tartu (1973)
15 R.I. Personov: in Spectroscopy and Excitation Dynamics of Condensed Molecular Systems, ed. by V.M. Agranovich, R.M. Hochstrasser (North-Holland, Amsterdam 1983) p. 555
16 M.U. Belyi, I.Ya. Kushnirenko, S.G. Nedel'ko: Ivz. Akad. Nauk SSSR Ser. Fiz. 43, 1133 (1979) [English Transl.: Bull. Acad. Sci. USSR Phys. Ser. 43, 28 (1979)]
17 I.Ya. Kushnirenko, V.R. Kumeskii: Opt. Spektrosk. 38, 928 (1975) [English transl.: Opt. Spectrosc. USSR 38, 534 (1975)]
18 K.K. Rebane, R.A. Avarmaa, L.A. Rebane: Izv. Akad. Nauk SSSR Ser. Fiz. 32, 1381 (1968)
19 L. Rebane, R. Avarmaa: Eesti NSV Tead. Akad. Toim. Füüs. Matem.(Proc. Estonian SSR Acad. Sci.) 17, 120 (1968) (in Russian)
20 L. Rebane, P. Saari, R. Avarmaa: Eesti NSV Tead. Akad. Toim. Füüs. Matem. (Proc. Estonian SSR Acad. Sci.) 19, 44 (1970) (in Russian)
21 K. Rebane, P. Saari: Eesti NSV Tead. Akad. Toim. Füüs. Matem.(Proc. Estonian SSR Acad. Sci.) 17, 241 (1968) (in Russian)
22 A. Freiberg, P. Saari: IEEE J. QE-19, 622 (1983)
23 A. Freiberg, J. Aaviksoo, K. Timpmann, J. Mol. Struct. 142, 563 (1986)
24 P. Saari: in Ultrafast Relaxation and Secondary Emission, ed. by O. Sild (Estonian SSR Acad. Sci., Tallinn 1979) p. 142
25 R. Avarmaa, L. Rebane: Phys. Status Solidi 35, 107 (1969)
26 A.R. Evans, D.B. Fitchen: Phys. Rev. B2, 1074 (1970)
27 R. Callender, P.S. Pershan: Phys. Rev. Lett. 23, 947 (1969)
28 T. Haldre, L. Rebane, O. Sild, E. Järvekülg: Eesti NSV Tead. Akad. Toim. Füüs. Matem.(Proc. Estonian SSR Acad. Sci.) 24, 417 (1975) (in Russian)
29 K.K. Rebane, L.A. Rebane, T.J. Haldre, A.A. Gorokhovskii: in Advances in Raman Spectroscopy, Vol. 1, Proc. 3rd Int. Conf. on Raman Spectroscopy, Reims, France, 1972, ed. by J.P. Mathieu (Heyden and Son, London 1973) p. 379
30 L.A. Rebane, A.M. Freiberg, J.J. Koni: Fiz. Tverd.Tela 15, 3318 (1973) [English transl.: Sov. Phys. - Solid State 15, 2209 (1974)]
31 A.B. Treshchalov, A.M. Freiberg, L.A. Rebane: Fiz. Tverd. Tela 17, 2816 (1975) [English transl.: Sov. Phys. - Solid State 17, 1884 (1975)]
32 A. Treshchalov, A. Freiberg, L. Rebane: Eesti NSV Tead. Akad. Toim. Füüs. Matem. (Proc. Estonian SSR Acad. Sci.) 24, 407 (1975) (in Russian)
33 L.A. Rebane, A.B. Treshchalov: Izv. Akad. Nauk SSSR Ser. Fiz. 40, 1926 (1976) [English transl.: Bull. Acad. Sci. USSR Phys. Ser. 40, 133 (1976)]
34 A.M. Freiberg, A.B. Treshchalov, O.I. Sild: Zh. Prikl. Spektrosk. 28, 808 (1978) [English transl.: J. Appl. Spectrosc. 28, 548 (1978)]

35 L.A. Rebane, A.B. Treshchalov, T.J. Haldre: Fiz. Tverd. Tela 16, 2236 (1974) [English transl.: Sov. Phys. Solid State 16, 1460 (1975)]
36 L. Rebane, A. Treshchalov: J. Lumin. 12/13, 425 (1976)
37 A.B. Treshchalov, L.A. Rebane: Fiz. Tverd. Tela 20, 469 (1978) [English transl.: Sov. Phys. - Solid State 20, 272 (1978)]
38 L.A. Rebane: in Luminescence of Inorganic Solids, ed. by B.D. Bartolo (Plenum Press, New York 1978) p. 665
39 A. Treshchalov: Eesti NSV Tead. Akad. Toim. Füüs. Matem.(Proc. Estonian SSR Acad. Sci.) 28, 233 (1979) (in Russian)
40 L. Rebane, P. Saari: Eesti NSV Tead. Akad. Toim. Füüs. Matem. (Proc. Estonian SSR Acad. Sci.) 19, 123 (1970) (in Russian)
41 L.A. Rebane: Opt. Spektrosk. 31, 230 (1971) [English transl.: Opt. Spectrosc. USSR 31, 123 (1971)]
42 L.A. Rebane, G.S. Zavt, K.E. Haller: Phys. Status Solidi b81, 57 (1977)
43 A.M. Stoneham: Rev. Mod. Phys. 11, 82 (1969)
44 A.M. Freiberg, L.A. Rebane: Fiz. Tverd. Tela 16, 2626 (1974) [English transl.: Sov. Phys. - Solid State 16, 1704 (1974)]
45 A. Freiberg, L.A. Rebane: in Molecular Spectroscopy of Dense Phases, ed. by M. Grosmann, S.G. Elkomoss, J. Ringeissen (Elsevier, Amsterdam 1976) p. 495
46 W. Voigt: S.B. bayer. Akad. Wiss. 1912, 603 (1912)
47 A.M. Freiberg: Ph.D. Thesis, Inst. Phys., Estonian SSR Acad. Sci., Tartu (1976)
48 A.M. Freiberg: Fiz. Tverd. Tela 17, 3400 (1975) [English transl.: Sov. Phys. - Solid State 17, 2224 (1975)]
49 R.H. Silsbee: Phys. Rev. 128, 1726 (1962)
50 D.E. McCumber, M.D. Sturge: J. Appl. Phys. 34, 1682 (1963)
51 M.A. Krivoglaz: Fiz. Tverd. Tela 6, 1707 (1964)
52 M.N. Sapozhnikov: Phys. Status Solidi B75, 11 (1976)
53 I.S. Osad'ko: Usp. Fiz. Nauk 128, 31 (1979) [English transl.: Sov. Phys. - Usp. 22, 311 (1979)]
54 L.A. Rebane: in Ultrafast Relaxation and Secondary Emission, ed. by O. Sild (Estonian SSR Acad. Sci., Tallinn 1979) p. 89
55 J. Friedrich, D. Haarer: Angew. Chem. 23, 113 (1984)
56 M.J. Weber (ed.): Optical Linewidth in Glasses, J. Lumin. 36, N 4/5 (1987)
57 A.A. Gorokhovskii: in this issue
58 D. Hsu, J.L. Skinner: J. Chem. Phys. 83, 2097 (1985); 83, 2107 (1985)
59 J.L. Skinner, D. Hsu: J. Phys. Chem. 90, 4931 (1986)
60 M.A. Krivoglaz: in this issue
61 V.V. Hizhnyakov: in this issue
62 L.A. Rebane, A.M. Freiberg: Izv. Akad. Nauk SSSR Ser. Fiz. 39, 1987 (1975) [English transl.: Bull. Acad. Sci. USSR Phys. Ser. 39, 179 (1975)]
63 A. Freiberg, L.A. Rebane: Phys. Status Solidi b81, 359 (1977)
64 V. Hizhnyakov, I. Tehver: Phys. Status Solidi b21, 755 (1967)
65 F. Luty: Cryst. Latt. Def. Amorph. Mat. 12, 343 (1985)
66 L.A. Rebane, O.I. Sild, T.J. Haldre: Izv. Akad. Nauk SSSR Ser. Fiz. 35, 1395 (1971)
67 A. Freiberg, L. Rebane: Eesti NSV Tead. Akad. Toim. Füüs. Matem. (Proc. Estonian SSR Acad. Sci.) 25, 380 (1976) (in Russian)
68 L. Rebane: Eesti NSV Tead. Akad. Toim. Füüs. Matem. Tehn. (Proc. Estonian SSR Acad. Sci.) 15, 301 (1966) (in Russian)
69 R.V. Jaaniso, R.A. Avarmaa: Zh. Prikl. Spektrosk. 44, 601 (1986)
70 R. Avarmaa, R. Jaaniso, K. Mauring, I. Renge, R. Tamkivi: Mol. Phys. 57, 605 (1986)
71 R. Avarmaa: in this issue
72 E. Realo: in this issue
73 R. Florian, L.O. Schwan, D. Schmid: Solid State Commun. 42, 55 (1982); Phys. Rev. A29, 2709 (1984)
74 Y. Yang, F. Luty: Phys. Rev. Lett. 51, 419 (1983)
75 W. Tkach, T.R. Gosnell, A.J. Sievers: Opt. Lett. 9, 122 (1984); Infrared Phys. 25, 35 (1985)
76 W.E. Moerner, A.J. Sievers, A.R. Shraplyvy: Phys. Rev. Lett. 47, 1082 (1981)
77 J.F. Pinto, L.W. Stratton, C.R. Pollock: Opt. Lett. 10, 384 (1985)

78 D. Wandt, W. Gellerman: Opt. Commun. 61, 405 (1987)
79 K.K. Rebane, L.A. Rebane: Izv. Akad. Nauk SSSR Ser. Fiz. 47, 1250 (1983) [English transl.: Bull. Acad. Sci. USSR Phys. Ser. 47, 1 (1983)]
80 A. Laisaar, A. Niilisk, A. Mugra: in High Pressure Science and Technology, Vol.2, ed. by B. Vodar, Ph. Marteau (Pergamon Press, Oxford 1980) p. 772

Zero-Phonon Lines in the Spectra of Polyatomic, Including Biogenic, Molecules

R. Avarmaa

Institute of Physics, Estonian SSR Acad. Sci.
SU-202400 Tartu, Estonian SSR, USSR

Abstract

A review of the investigation of the zero-phonon lines and vibronic structure in the spectra of frozen solutions of chlorophyll and related pigments as well as of some native systems is given. The methods of selective low-temperature laser spectroscopy have been used. Theoretical aspects of the influence of the inhomogeneous structure of the matrix and energy transfer on spectral lines are considered.

1. Introduction

Since its foundation the theory of zero-phonon lines (ZPL) has been used [1] for the interpretation of Shpol'skii spectra [2], i.e. quasi-line spectra of organic molecules embedded in specific (mainly n-alcane) solvent matrices. The zero-phonon nature of the observed lines has been reliably verified by a selective excitation of individual components of the fluorescence spectral multiplet and by the observation of phonon wings in the respective excitation [3] as well as fluorescence spectra [4]. Several studies of thermal broadening and shift of ZPL in Shpol'skii spectra have been conducted [5,6], which have generally confirmed the results of the theory [7]. Also, it has been theoretically estimated [7] that purely electronic lines (PEL) at sufficiently low temperatures remain inhomogeneously broadened (of the order of 1 cm^{-1}).

In parallel to this direction, studies on ZPL in inorganic crystals appeared and an appropriate theory was developed [8-11]. Here we mention the experimental studies on rare-earth impurity ions in fluoride crystals [12,13], the observation of PEL as an optical analog of the Mössbauer effect [14] in a CdS crystal [15] as well as in colour centres in alkali halides [16]. An essential contribution to the understanding of the processes of electron-vibrational interactions of the impurity-crystal system has been made by the investigation of ZPL in the spectra of small molecular ions in alkali halides (see, e.g. [17,18]). Thus, quantitative data about homogeneous and inhomogeneous broadening and on the role of pseudo-local vibrations have been obtained [19].

In the early seventies, it was clear that inhomogeneous broadening is the main factor impeding the obtaining and investigating narrower PEL on the lowering of the temperature. In an earlier attempt to exclude inhomogeneous broadening, a detectable narrowing of the fluorescence line of an europium impurity was achieved via resonant lamp excitation [20]. Laser resonant excitation of a ruby crystal has lead to a remarkable narrowing of PEL (up to 0.007 cm^{-1}), inhomogeneous broadening being almost fully eliminated [21]. Non-resonant laser excitation of rare-earth ions doped into inorganic glasses has resulted only in moderate fluorescence line narrowing (FLN) [22].

As to organic molecules in not specially-chosen matrices (mainly amorphous ones), the situation was more complicated: spectra of the majority of frozen solutions remained practically structureless even at liquid helium temperatures. Selective laser excitation has lead to a striking effect of the transformation of such broad bands into spectra with narrow vibronic lines as a result of considerable elimination of inhomogeneous broadening [23] (this technique is now called site-selection spectroscopy). Again, the presence of narrow intense ZPL in homogeneous spectra is a necessary prerequisite for the appearance of the effect.

In case of such complex, biologically important molecules as chlorophyll (Chl) (Figure 1) and its analogs there was a hope to obtain fine-structured spectra with the aid of the Shpol'skii effect, but its applicability to these molecules turned out to be rather limited [24,25]. Under laser excitation some narrow lines have been detected in the fluorescence spectra of Chl-a in amorphous matrices at T = 4.2 K [26], in analogy with simpler molecules [23]. The interpretation of the observed complicated structure of the spectra have stimulated a special theoretical treatment, as a result of which some general relations for the widths and intensities of lines in inhomogeneous spectra have been obtained [27]. Also, an important conclusion about the absence of any correlation between the line structure of different electronic states

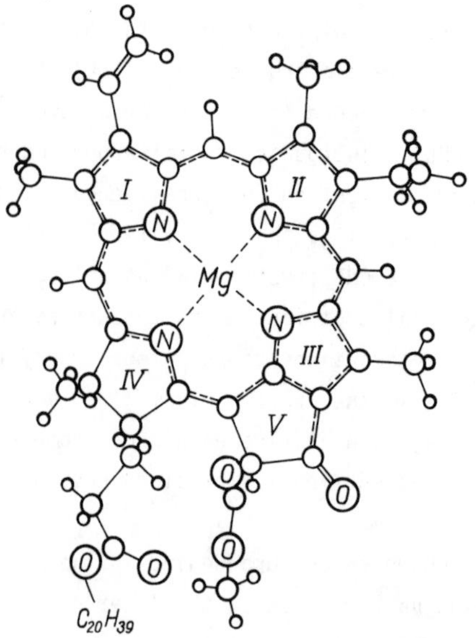

Figure 1: The structure of a chlorophyll-a molecule. Small circles - hydrogen atoms, medium circles - carbon, the rest of the symbols are indicated in the Figure

of solute molecules has been drawn [28], and a double scanning method introduced [29].

One of the main conclusions of the theory [27] is that under nonresonant selective excitation the observed PEL obtains the width of the corresponding vibronic absorption line, i.e. retains certain part of inhomogeneous broadening. In view of difficulties with on-resonance fluorescence measurements, a method of spectral hole burning has been proposed as a means for measuring homogeneous contours of PEL [30, 31]. Later, a number of other application fields have been found for this method (see some other articles in this issue). It has proved to be very useful in case of chlorophyll-like molecules as well.

2. Theoretical Treatment of Selectively-Excited Spectra

2.1. Width, Intensity and Frequency of ZPL

The effect of selective excitation of inhomogeneously-broadened spectra consists in the narrowness and large relative intensity of ZPL (especially of PEL) in the corresponding homogeneous spectra [32]. For the treatment of site selection spectra the introduction (independently in several papers [33-35]) of a concept of inhomogeneous distribution function (IDF), expressing the distribution of the number of solute molecules over the shifts $\Delta\nu$ of the purely electronic transition, has turned out to be quite useful. As a result, a monochromatically-excited fluorescence spectrum $F(\nu_e,\nu_f)$ can be presented as a convolution of the IDF $\rho(\Delta\nu)$ with homogeneous absorption and fluorescence spectra. Main conclusions [27] are the following: (i) the observable linewidth is formed by summing up the respective homogeneous widths of the emission and absorption lines; (ii) in case of large inhomogeneous broadening in the 0-0 band region several pseudo-lines are found (they are all PEL by their origin), whose distances from the excitation frequency are equal to the local (intramolecular) vibrational frequencies in the excited electronic state; (iii) relative intensity of the lines depends essentially on the frequency of excitation within IDF.

A pseudoline structure was experimentally found in the spectra of tetracene dissolved in Shpol'skii matrices (Figure 2). It should be pointed out that the frequencies determined from such spectra correspond to the vibrations of solute molecules in the excited state (S_1), in contrast to the case of excitation on 0-0 transition when ground state (S_0) frequencies appear.

In several subsequent papers, the methods for obtaining IDF and homogeneous contours from experimental spectra have been elaborated [29,36-38] and model calculations, quantitatively describing the diminishing of the ZPL intensity in the emission spectrum on the short-wave displacement of the excitation frequency, have been performed [39,40].

In this connection it is useful to present expressions [41] for the effective values of the Debye-Waller factor (DWF), β_f and β_e, measurable from the fluorescence

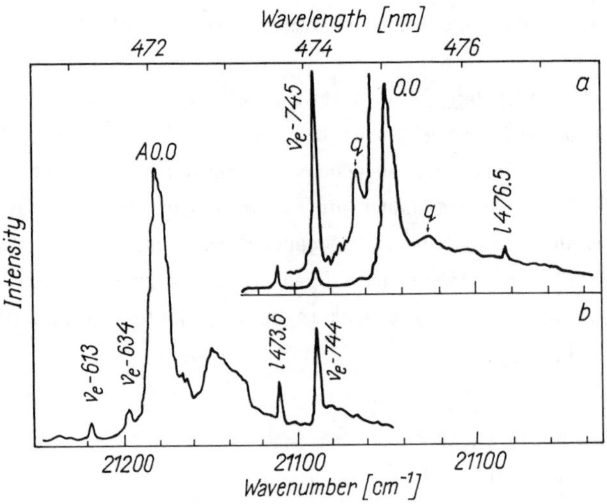

Figure 2: The fluorescence spectra of tetracene in hexane (a) and nonane (b) under laser excitation at 457.9 nm, T = 4.2 K

and excitation spectra as functions of the excitation (ν_e) or recording (ν_f) frequencies, respectively:

$$\beta_f(\nu_e) = \beta^2 \rho(\Delta\nu_e)/\bar{\kappa}(\nu_e) \;, \tag{1}$$

$$\beta_e(\nu_f) = \beta^2 \rho(\Delta\nu_f)/F(\nu_f) \;, \tag{2}$$

where $\bar{\kappa}(\nu_e)$ is inhomogeneously-broadened absorption spectrum, $F(\nu_f)$ is the corresponding fluorescence spectrum at "white" excitation and β, the DWF value for the homogeneous spectrum. As a result of inhomogeneous broadening, in general, the mirror symmetry between the frequency-selected fluorescence and excitation spectra does not hold. However, if the homogeneous subspectra $\kappa(\nu_e)$ and $\phi(\nu_f)$ are mirror-symmetric with respect to one another, and if the IDF $\rho(\Delta\nu)$ is a symmetric curve, we get a new two-dimensional symmetry between the fluorescence and excitation spectra:

$$F(\nu_e = \nu', \nu) = \varepsilon(-\nu, \nu_f = \nu') \;, \tag{3}$$

where $\varepsilon(\nu, \nu_f)$ is an excitation spectrum for the recording frequency ν_f, whereby the frequencies are counted from the unshifted ($\Delta\nu = 0$) PEL. From (3) follows also a simple relation between the DWF in the corresponding spectra:

$$\beta_f(-\nu') = \beta_e(\nu') \;. \tag{4}$$

Relation (4) is in accordance with the conclusions [39,40] that the ZPL intensity in emission spectra is growing when the excitation is red-shifted ($\nu' < 0$), while in fluorescence spectra it is increasing with the blue-shift of the recording frequency ($\nu' > 0$).

2.2. Inclusion of Energy Transfer

The preceding calculations were carried out for the case of low concentration, when the transfer of excitation energy between the solute molecules can be neglected. In connection with the studies of the spectra of Chl in model and native systems it was necessary to consider the problem of site-seletion spectra in case of fast energy transfer. A treatment of such situation has been given in [41].

Let us start from the model of Sect. 2.1 and, in addition, take into account the transfer of energy between the impurity molecules within the inhomogeneous manifold by using the Förster dipole-dipole approximation [42]. It is suitable to consider a function of energy transfer, $p(\nu_d)$, which gives the average probability of transfer from a certain donor to all possible acceptors within the inhomogeneous band. It should be pointed out that the nonresonant transfer, taking place as a result of the overlap of the phonon wings of the donor emission spectrum and the acceptor absorption spectrum, leads to the disappearance of ZPL.

Supposing that the averaging procedure can be performed independently over the distances and frequencies, we get

$$p(\nu_d) \sim \int d\nu_a \rho(\nu_a) S(\nu_d - \nu_a) \ , \tag{5}$$

where the overlap integral $S(\nu_d - \nu_a)$ determines the transfer radius R_0:

$$R_0^6 \sim S(\nu_d - \nu_a) = \int d\nu_t \rho(\nu_t - \nu_d) \kappa(\nu_t - \nu_a) \ , \tag{6}$$

ν_a is the 0-0 transition frequency of an acceptor. Recently, it has been shown [43] that a strict averaging results in a modified transfer function:

$$q(\nu_d) \sim \int d\nu_a \rho(\nu_a) \sqrt{S(\nu_d - \nu_a)} \ . \tag{7}$$

In Figure 3,we present the results of the model calculations [41] of the functions p and q, which express an overall tendency of a decrease of the transfer rate on the low-frequency slope of the inhomogeneous band. This is due to the decrease of the number of appropriate (in the sense of the value of overlap integral (6)) acceptors

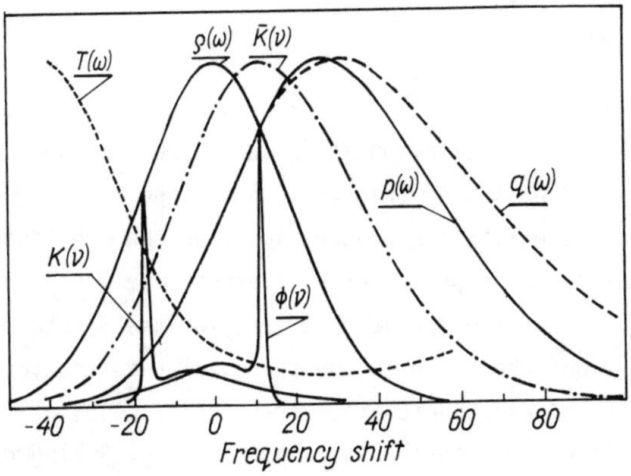

Figure 3: The inhomogeneous distribution function $\rho(\omega)$ and the functions of energy transfer $p(\omega)$, $q(\omega)$, calculated for T = 0 by using the homogeneous model spectra $k(\nu)$ and $\phi(\nu)$

on the lowering of ν_d. There is no great difference between the functions p and q, so a simpler expression (5) can be used in practice. In this case, the decrease of the relative intensity of ZPL is expressed by the frequency dependence of the donor fluorescence yield, $T(\omega)$ (Figure 3).

The actual shape of the fluorescence spectrum was calculated on the basis of the numerical solution of the integral equation with respect to $f(\nu_e,\Delta\nu)$ - the inhomogeneous spectral distribution of the excited state population - on the assumption of monochromatic excitation. Now the fluorescence spectrum is given by

$$F_t(\nu_e,\nu_f) = \int d(\Delta\nu) f(\nu_e,\Delta\nu)\phi(\nu_f-\Delta\nu) \quad . \qquad (8)$$

The results of model calculations showed that, in spite of the considerable reduction of the ZPL intensity in the centre of the inhomogeneous band, it is possible to obtain quite distinct lines in the low-frequency edge even if the concentration is near the critical, $C \approx \langle R_0\rangle^{-3}$. From the treatment of the conjugated excitation spectra it can be concluded that relations (1) - (4) are violated as the result of the energy transfer. At the same time the frequency dependence of the DWF for fluorescence spectra increases, contrary to the excitation spectra, where it is smoothed out. In this sense, ZPL in excitation spectra are less sensitive to the energy transfer than in fluorescence spectra, where the lines quickly vanish in the high-frequncy region of IDF.

Since these calculations have been performed for the resonant 0-0 group, in a real experiment, where excitation takes place in the vibronic absorption band, the overlapping of numerous (pseudo-) lines and the corresponding wings in the purely electronic region must be taken into account. It can be concluded that the most appropriate way to get a ZPL is to measure the excitation spectrum, while fluorescence detection has been set at the blue edge of the 0-0 band.

Note that the hole-burning process is also distorted by the energy transfer: now the relative efficiency of the hole burning is given by the excitation function $f(\nu_e,\Delta\nu)$ which is different from the absorption spectrum.

3. High-Resolution Spectra of Chlorophyll-Like Molecules

3.1. Site-Selection Spectra of Frozen Solutions

In the fluorescence spectrum of Chl-a at liquid helium temperatures, a typical pseudoline structure arises under the He-Ne laser excitation at 632.8 nm [26]. With the rise of temperature the ZPL broadening and weakening (Figure 4) predicted by the theory are observed [7]. By use of various excitation frequencies from the dye laser similar sharp-line spectra for a number of chlorophyll derivatives have been obtained (e.g. Figure 5) and the frequencies of the vibrational sets for the S_1-state have been determined. Excitation in 0-0 band has enabled one to reveal the structure of the fluorescence spectrum for the ground electronic state frequencies [44], although these measurements are more difficult due to the fast hole burning. Analogous

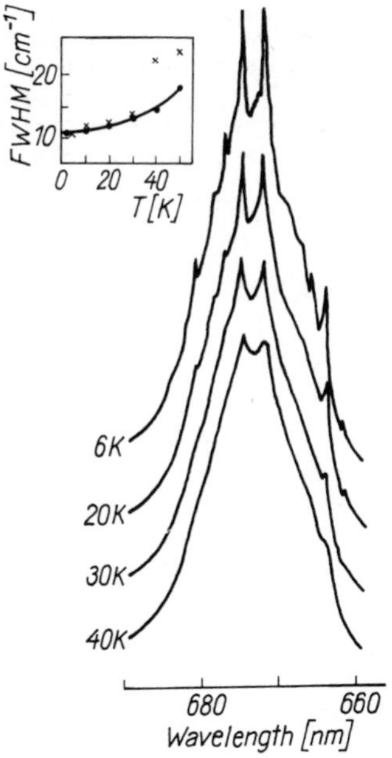

Figure 4: The fluorescence spectra of Chl-a in ether at different temperatures, $\lambda_e = 632.8$ nm. Inset: temperature dependence of the width of 664 nm (o) and 674.8 nm (x) lines

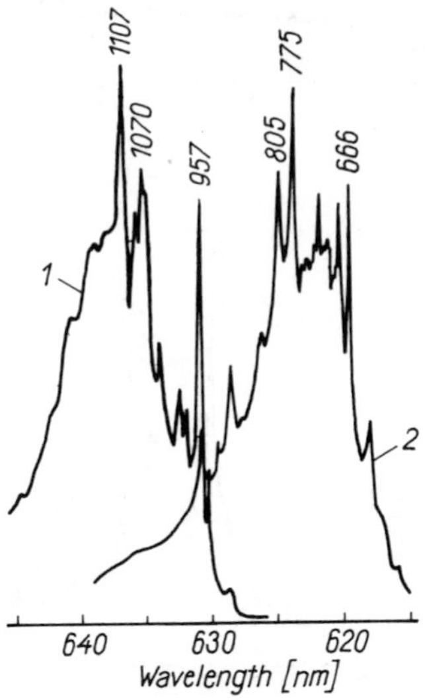

Figure 5: The fluorescence spectra of mono- (2) and disolvates (1) of protochlorophyll under a selective excitation ($\lambda_e = 596$ nm). Solvents: diethyl ether and mixture of ether and butanol, respectively; T = 5 K

results on the fluorescence excitation spectra of porphyrins [45,46] and Chl [47,48] have been obtained in other laboratories.

Vibrational frequencies and intensities of $S_1 \leftarrow S_0$ transitions become apparent in the most natural manner in the excitation spectra, when the narrow-band recording of the fluorescence is performed on the purely electronic transition. Such kind of experiments have become feasible thanks to the application of tunable dye lasers with a constant output power. An exact calibration of the wavelength scale by an automatic optogalvanic system [49] is also of importance.

In Figure 6, a sharp-line excitation spectrum of a metal-free chlorophyll, pheophytin-a, is shown alongside with the polarization spectrum which exhibits some negatively-polarized lines. Besides, broad minima have been found in the polarization spectra of Chl-a and protochlorophyll (PChl), which have made possible a positive fixing of the $S_2 \leftarrow S_0$ band positions [50]. The energy difference, $S_2 - S_1$, is rather small, about 400 cm^{-1} for PChl and 800 - 900 cm^{-1} for Chl-a. The smallness of the splitting of the excited electronic state, degenerated in the zero approximation, indicates the symmetry of the central part of these molecules, which is in contrast to pheophytin

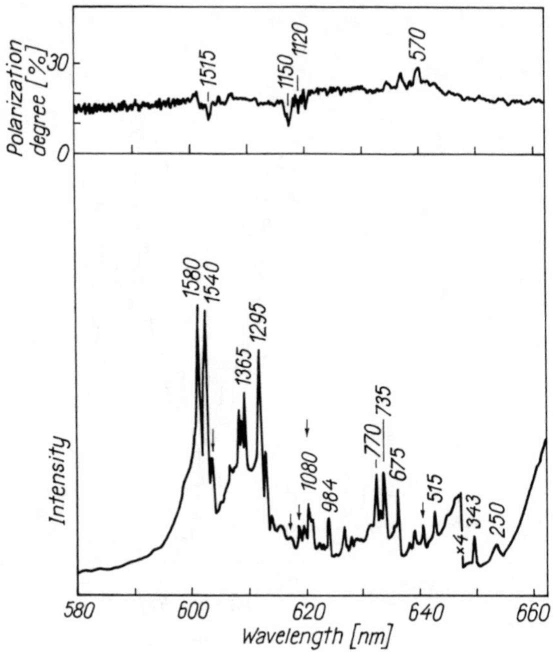

Figure 6: The fluorescence excitation spectrum of pheophytin-a in a glassy matrix for λ_f = 665 nm (lower curve) and the corresponding polarization spectrum (upper curve)

and H_2-porphyrins, where the molecules' centres possess the second-order symmetry axis, the $S_2 - S_1$ distance being therefore much longer, more than 3000 cm^{-1}.

Detailed data on the site-selection spectra of Chl and its analogs, as well as the interpretation of the vibrational structure are presented in a recent review [51] (see also [46,52,53]).

3.2. Hole Burning

The process of photoburning of persistent spectral holes starts with a selective excitation of an impurity molecule, to which a photoreaction follows. As a result, with a certain probability, the molecule does not return to its initial state. The reaction may be photochemical - then the spectrum of the photoproduct is usually strongly shifted from the initial position. The so-called nonphotochemical mechanism, when the interaction between the impurity and the host is changed as a result of photoexcitation, is often found to operate in glassy matrices. A purely photophysical hole-burning method, consisting in the saturation of the $S_1 \leftarrow S_0$ transition at the expense of the triplet state population, has been proposed [54] to be applied to Chl. This method has enabled one to reveal [55] narrow ZPL and the corresponding phonon wing in the spectrum of PChl (Figure 7).

In the course of these measurements a slow phase in the hole burning and hole recovery processes, assigned to a nonphotochemical mechanism, was observed besides the triplet state kinetics. The persistent nature of such holes enabled their thorough

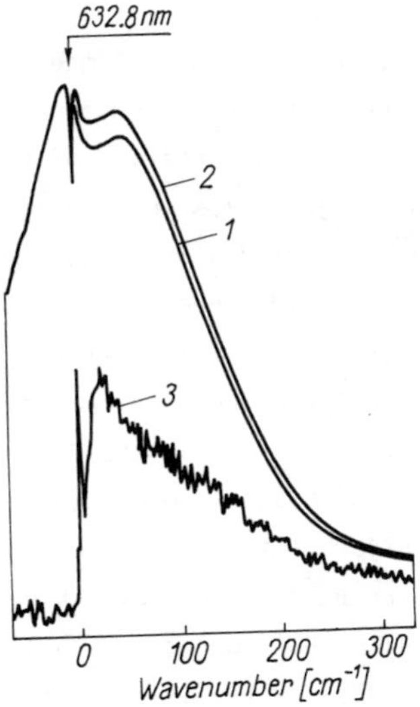

Figure 7: A metastable hole in the fluorescence spectrum of protochlorophyll, measured without delay (1) and with a delay of 15 ms (2); 3 - the difference spectrum

investigation. In particular, the holewidth of 300 MHz was obtained at 1.8 K, i.e. the record resolution in Chl spectroscopy was achieved. One managed to elucidate the mechanism of the burning-recovery process [56]. Firstly, it turned out that the frequencies of the burnt-out impurities are uniformly distributed over the inhomogeneous band, proportionally to the initial IDF. Secondly, the irreversability of the thermal hole-filling on cycling temperature changes showed that every impurity molecule has a large number of impurity-host configurations that cover an over 100 cm^{-1} interval in the energy scale.

A more efficient photochemical mechanism (quantum yield $\sim 10^{-3}$) has been found in case of pheophytin [57], which consists in the rotation of central protons and which has earlier been known for porphyrins [30,58]. Because of the molecule's skeleton asymmetry, the $S_1 - S_0$ transition energies of the two tautomers differ more drastically, ~ 500 cm^{-1}. The photoproduct is rather unstable, the probability of its photodecomposition being two orders of magnitude higher than that of the formation, while the barrier height for thermal decomposition makes only ≈ 50 cm^{-1}. Recently, analogous quasistable photochemical tautomers were discovered also for other porphyrin molecules with isocycle [59,60].

The existence of the efficient photochemical burning mechanism has facilitated the determination of the homogeneous widths of the vibronic levels of pheophytin. For this purpose holes were burnt in the vibronic lines of the excitation spectrum (cf. Figure 6) and difference spectra were calculated (see below). In this way, homogeneous widths and the corresponding lifetimes at about 30 vibrational sublevels of S_1-state were found for pheophytin [61]. Relaxation times were found to be within the 1 - 6 ps range, with the tendency of getting shorter at higher frequencies. Thus, the complexity of the molecule's peripheral structure as compared to simple porphyrins does not cause any noticeable growth of the vibrational relaxation rate.

The combining of the methods of selective excitation and hole burning has made it possible to develop a fine, experimentally simple method for determining homogeneous spectra and IDF [38]. The tremendous difference between the peak intensities of the ZPL and the phonon wing, mentioned above, is used here, which justifies the neglection of hole burning through the wings at the initial phase of irradiation. The homogeneous fluorescence spectrum is produced in the following way. The recording of the spectra is performed twice under a very weak monochromatic excitation, so that the hole burning during the measurements can be ignored. Between these two measurements the hole burning is performed at the very excitation frequency, ν_e. It is the difference of these two spectra that gives the homogeneous emission spectrum of the impurities whose ZPL lie at the frequency ν_e, except the contour of the resonance line itself. Analogously, the homogeneous absorption spectrum can be obtained, by making use of two site-selection excitation spectra recorded with the hole burnt meanwhile at the registration frequency. It should be noted that if, besides the shift of the electronic transition frequency (as in the basic model [33]), the inhomogeneity includes also the variations of the electron-vibrational interaction, then the spectra obtained are averaged over such characteristics as DWF and the structure of the phonon wing. To reveal the IDF it is necessary to measure also the inhomogeneous spectrum (either fluorescence or excitation spectrum), which from mathematical point of view represents a convolution of the corresponding homogeneous spectrum and IDF. In [38], a convenient algorithm for the deconvolution and finding IDF is presented.

The method described was used [62] to determine these homogeneous characteristics of PChl as impurity molecules in ether/butanol glass for ten ν_e values. The results of measurements indicate the necessity of taking into account the discrepancies from the simple model which assumes linear electron-vibrational interaction (Figure 8). The shape of the phonon wing in fluorescence spectra proved to be practically unchanged on excitation within the inhomogeneous band. This confirms the common assumption [33] that all the impurity molecules have homogeneous spectra of the same shape. For the conjugated transition $S_1 \leftarrow S_0$, as can be seen in Figure 8, such simple picture does not hold: for various ν_{oo} homogeneous wings differ considerably. Another pecularity is the deviation from the mirror symmetry of the absorption and excitation spectra - in absorption the interaction with phonons is stronger. It is possible to indicate several mechanisms which can cause the difference in phonon wings in absorption and emission, such as the deviation from the Condon approximation, the mixing of the normal coordinates and nonadiabatic interaction. In the present case, the consideration of nonadiabatic interaction is especially relevant because of the small $S_2 - S_1$ splitting, which can additionally cause the violation of the Condon approximation. The fact that the shape of the phonon wing in the absorption spectrum undergoes a change within the IDF (contrary to the emission spectrum) is also explained by the perturbing role of S_2-state, since S_2-S_1 splitting depends on $\Delta\nu$ [50].

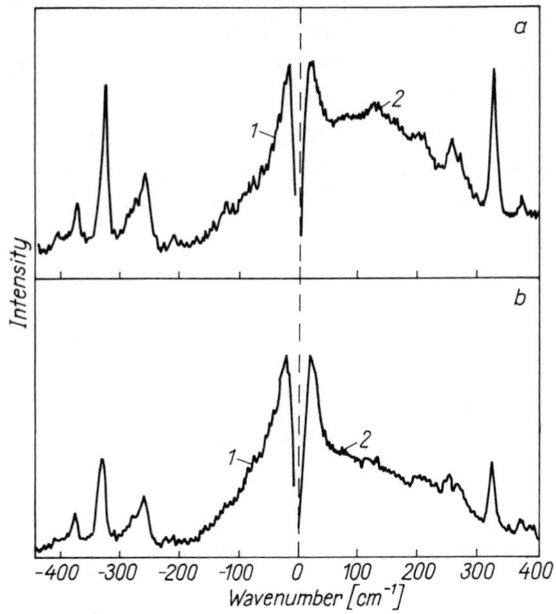

Figure 8: Homogeneous fluorescence (1) and absorption (2) spectra of protochlorophyll molecules in ether/butanol glass for 0-0 transition frequencies of molecules at 15782 cm^{-1} (a) and 15640 cm^{-1} (b)

The IDF obtained for this system is a smooth slightly asymmetric curve with FWHM = = 175 cm^{-1}.

3.3 Narrow Lines in the Spectra of Native Systems

The attempts to observe the fine vibronic structure in fluorescence spectra of green leaves and chloroplasts at selective excitation have not proved successful. In the course of time we started to suspect that the absence of selectivity arises mainly from one of the several possible reasons [63], namely, from the energy migration.

On the other hand, the hole-burning method has successfully been applied to the investigation of electron transfer in the reaction centres of the photosystem I P700 [64] and to the primary process in photosystem II [65]. The primary charge separation time (tens of picoseconds) was evaluated from the width of persistent holes. The hole burning was used to study native pigment-protein complexes, such as phycobiliproteins [66]. Site-selection fluorescence spectra have been obtained for another complex, the metal-free cytochrome C [67]. These results have demonstrated that pigment-protein interactions do not always reduce drastically the intensity of ZPL.

The possible role of the energy transfer rate as a main selectivity-destroying factor (theoretical reasons for that, see Sect. 2.2) was substantiated in a study of etiolated (grown in the darkness) leaves [68]. The concentration of pigments in etiolated leaves is rather low, which excludes an efficient energy migration. Before exposition to the light the leaves contain two protochlorophyllide forms - the inactive

P630 and the active P650. Under selective excitation a pseudo-sharp-line structure can be seen in the fluorescence spectrum of a non-active form (Figure 9), while the S_1-state vibrational frequencies are close to those of monosolvated protochlorophyll in solutions [53]. A similar vibronic structure was observed in the excitation spectra on narrow-band detection. However, on vibronic excitation the sharp-line structure for P650 was not observable. This may be connected with a tight pigment packing in this protein complex.

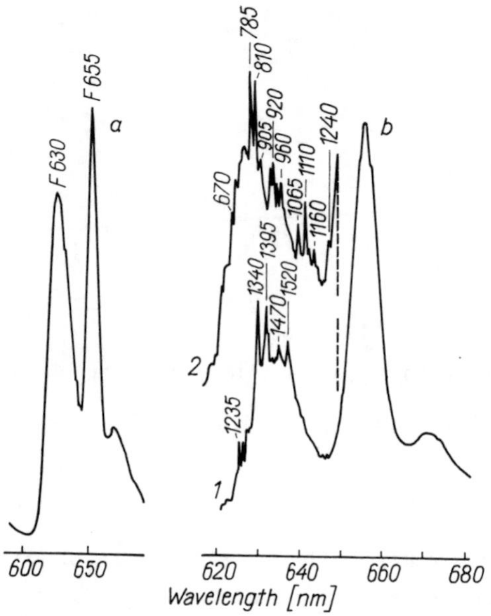

Figure 9: The fluorescence spectra of an etiolated barley leaf under excitation at 441.6 nm (a) and under a selective excitation (b) at 580.9 nm (1) and 599.1 nm (2). Vibrational frequencies are shown

At the initial stages of illumination at physiological temperatures P650 is converted into chlorophyllide, which turns into Chl after passing several transient forms. When the recording was set at the blue edge of the emission band, the vibronic sharp-line structure was revealed in the excitation spectrum of chlorophyllide in greening samples [68]. The shift of ν_f towards the band maximum or the accumulation of the pigment after a prolonged illumination lead to the disappearance of the spectral structure, which is in accordance with the conclusions about the influence of energy transfer on the ZPL intensities within an inhomogeneous system, drawn in Sect. 2.2.

Since the absence of lines in the spectrum of $S_1 \leftarrow S_0$ vibronic transitions may be brought about by the enhanced vibronic mixing of states in a dimeric (or associated) pigment, the possibility of observing ZPL in the 0-0 band of a photoactive form still remained. This was realized by using the hole-burning technique [69]. After a moderately-intensive exposition (tens of mW·cm^{-2} during some minutes) with a monochromatic light a stable narrow hole in P650 band was formed (Figure 10). A repeated burn-

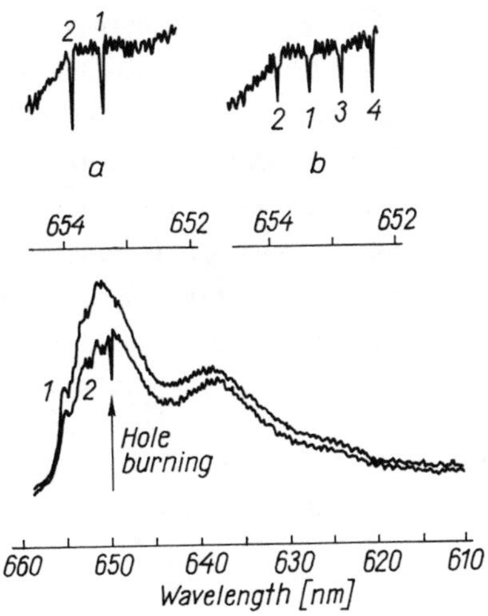

Figure 10: The fluorescence excitation spectra of an etiolated leaf at T = 5 K before (1) and after (2) hole burning at 649.05 nm (lower) and after a repeated hole burning (upper) in the sequence from 1 to 4

ing nearby did not lead to the filling of the existing holes, except some dark-filling. Thus, mainly irreversible photochemical transformations prevail in burning processes. It is possible that a non-fluorescent intermediate of the protochlorophyllide reduction is formed [70].

Consequently, the existence of ZPL in a photoactive protochlorophyllide-protein complex has been proved unequivocally. Furthermore, in the fluorescence spectrum a sharp-line structure of this form has been found [69]. Recently it was demonstrated in our laboratory that narrow (<1 cm^{-1}) persistent spectral holes can be burnt in the long-wavelength antenna forms of bacteriochlorophyll-c in green sulfur bacterium *Chlorobium limicola* as well as Chl-a in green post-etiolated maize leaves (in the spectral region of 700 - 745 nm) [74]. Further studies of the widths and shifts of ZPL as a function of temperature and external fields should give valuable information about the nature and strength of pigment-pigment and pigment-host interactions.

4. Summary

A series of studies have been carried out, leading to a conclusion that chlorophyll-type molecules have as good (i.e. narrow and intensive) ZPL as many other organic molecules do. An extensive cooling in a solid matrix or in a supersonic jet is necessary for these lines to become manifest. The latter technique has been applied to some porphyrins [72], but Chl, not to speak of its native complexes, is decomposed before the appropriate vapour pressure is achieved. Besides the cooling up to low temperatures, it is necessary to get rid of inhomogeneous broadening inherent to many solid systems. For this purpose the methods of selective laser excitation of molecules either in solid hosts or in a native environment are used.

Further progress in this field could be achieved, first of all, by studying different systems. For example, the mechanisms of photosynthesis in bacteria have intensively been studied by picosecond techniques [73], a number of problems in them being therefore now better understood than in two-system photosynthesis of higher plants. However, up to now there is no evidence about sharp-line spectra of bacteriochlorophyll-a. Recently, the methods of selective spectroscopy were applied to bacteriochlorophyll-a in our laboratory [75]. The studies of highly-resolved spectra of dimers and pigment-protein complexes as model systems of photosynthesis should also be of considerable importance.

I should like to express my gratitude to Prof. K.Rebane for his continuous interest in the present trend of investigations as well as to my co-authors for fruitful collaboration.

References

1 K.K. Rebane, V.V. Hizhnyakov: Opt. Spektrosk. 14, 362 (1963) [English transl.: Opt. Spectrosc. USSR 14, 193 (1963)]
2 E. Shpol'skii: Usp. Fiz. Nauk 80, 255 (1963) [English transl.: Sov. Phys.-Usp. 80, 411 (1963)]
3 K. Rebane, P. Saari, T. Tamm: Eesti NSV Tead. Akad. Toim. Füüs. Matem. (Proc. Estonian SSR Acad. Sci.) 19, 251 (1970) (in Russian)
4 R.I. Personov, I.S. Osad'ko, E.D. Godyaev, E.I. Al'shits: Fiz. Tverd. Tela 13, 2653 (1971)
5 R.I. Personov, V.V. Solodunov: Fiz. Tverd. Tela 10, 1848 (1968)
6 M.N. Sapozhnikov: Fiz. Tverd. Tela 15, 3160 (1973) [English transl.: Sov. Phys.-Solid State 15, 2111 (1974)]
7 K.K. Rebane: Elementarnaya teoriya kolebatel'noj struktury spektrov primesnykh tsentrov kristallov (Nauka, Moskva 1968) [English transl.: Impurity Spectra of Solids (Plenum Press, New York, 1970)]
8 J.J. Hopfield: in Proc. Intern. Conf. Semicond. Phys. (Exeter 1962) p. 75
9 R. Silsbee: Phys. Rev. 128, 1726 (1962)
10 M.A. Krivoglaz: Fiz. Tverd. Tela 6, 1707 (1964) [English transl.: Sov. Phys. - Solid State 6, 1340 (1964)]
11 D.E. McCumber, M.D. Sturge: J. Appl. Phys. 34, 1682 (1963)
12 P.P. Feofilov, A.A. Kaplyanskii: Opt. Spektrosk. 12, 493 (1962) [English transl.: Opt. Spectrosc. USSR 12, 272 (1962)]
13 W. Kaiser, C.G.B. Garett, D.L. Wood: Phys. Rev. 123, 766 (1961)
14 E.D. Trifonov: Dokl. Akad. Nauk SSSR 147, 826 (1962)
15 E.F. Gross, S.A. Permogorov, B.S. Razbirin: Dokl. Akad. Nauk SSSR 154, 1306 (1964)
16 D.B. Fitchen, R.H. Silsbee, T.A. Fulton, E.L. Wolf: Phys. Rev. Lett. 11, 275 (1963)
17 Molekulyarnye tsentry svecheniya O_2^- i S_2^- v shchelochnogalojdnykh kristallakh, Trudy IFA AN ESSR, Vol. 37 (Valgus, Tallinn 1968)
18 K.K. Rebane, L.A.Rebane: Pure and Appl. Chem. 37, 161 (1974)
19 A. Freiberg, L.A. Rebane: in this issue
20 Yu.V. Denisov, V.A. Kizel: Opt. Spektrosk.: 23, 472 (1967) [English transl.: Opt. Spectrosc. USSR 23, 251 (1967)]
21 A. Szabo: Phys. Rev. Lett. 25, 924 (1970)
22 L.A. Rieseberg: Phys. Rev. Lett. 28, 786 (1972)
23 R.I. Personov, E.I. Al'shits, L.A. Bykovskaya: Pis'ma Zh. Eksp. Teor. Fiz. 15, 609 (1972) [English transl.: JETP Lett. 15, 431 (1972)]
24 F.F. Litvin, R.I. Personov, O.N. Korotaev: Dokl. Akad. Nauk SSSR 188, 1169 (1969)

25 R. Avarmaa, K. Rebane: Eesti NSV Tead. Akad. Toim. Füüs. Matem. (Proc. Estonian SSR Acad. Sci.) 22, 108 (1973) (in Russian)
26 R. Avarmaa: Eesti NSV Tead. Akad. Toim. Füüs. Matem. (Proc. Estonian SSR Acad. Sci.) 23, 93 (1974) (in Russian)
27 K.K. Rebane, R.A. Avarmaa, A.A. Gorokhovskii: Izv. AN SSSR Ser. Fiz. 39, 1793 (1975) [English transl.: Bull. Acad. Sci. USSR Phys. Ser. 39, 8 (1975)]
28 P.M. Saari, T.B. Tamm: Izv. AN SSSR Ser. Fiz. 39, 2321 (1975) [English transl.: Bull Acad. Sci. USSR Phys. Ser. 39, 80 (1975)]
29 T.B. Tamm, J.V. Kikas, A.E. Sirk: Zh. Prikl. Spektrosk. 14, 315 (1976)
30 A.A. Gorokhovskii, R.K. Kaarli, L.A. Rebane: Pis'ma Zh. Eksp. Teor. Fiz. 20, 474 (1974)
31 V.M. Kharlamov, R.I. Personov, L.A. Bykovskaya: Opt. Commun. 12, 191 (1974)
32 K.K. Rebane: Zh. Prikl. Spektrosk. 37, 906 (1982) [English transl.: J. Appl. Spectrosc. 37, 1346 (1982)]
33 R. Avarmaa: Eesti NSV Tead. Akad. Toim. Füüs. Matem. (Proc. Estonian SSR Acad. Sci.) 23, 238 (1974) (in Russian)
34 T. Kushida, E. Takushi: Phys Rev. B12, 824 (1975)
35 W.C. McColgin, A.P. Marchetti, J.H. Eberly: J. Am. Chem. Soc. 100, 5622 (1978)
36 J. Kikas: Eesti NSV Tead. Akad. Toim. Füüs. Matem. (Proc. Estonian SSR Acad. Sci.) 25, 374 (1976) (in Russian)
37 J. Fünfschilling, I. Zschokke-Gränacher, D.E. Williams: J. Chem. Phys. 75, 3669 (1981)
38 R. Jaaniso: Eesti NSV Tead. Akad. Toim. Füüs. Matem. (Proc. Estonian SSR Acad. Sci) 34, 277 (1985) (in Russian)
39 J. Kikas: Chem. Phys. Lett. 57, 511 (1978)
40 M.N. Sapozhnikov, V.I. Alekseev: Chem. Phys. Lett. 97, 331 (1983)
41 R. Avarmaa, R. Jaaniso, K. Mauring, I. Renge, R. Tamkivi: Molec. Phys. 57, 605 (1986)
42 V.M. Agranovich, M.D. Galanin: Perenos energii elektronnogo vozhbuzhdeniya v kondensirovannykh sredakh (Nauka, Moskva 1978)
43 R.A. Avarmaa, R.V. Jaaniso: Fiz. Tverd. Tela 29, 1832 (1987)
44 L.A. Bykovskaya, F.F. Litvin, R.I. Personov, Yu.V. Romanovskii: Biofizika 25, 13 (1980)
45 Yu.V. Romanovskii: Eesti NSV Tead. Akad. Toim. Füüs. Matem. (Proc. Estonian SSR Acad. Sci.) 31, 139 (1982) (in Russian)
46 K.N. Solov'ev, L.L. Gladkov, A.S. Starukhin, S.F. Shkirman: Spektroskopiya porfirinov: kolebatel'nye sostoyaniya (Nauka i Tekhnika, Minsk 1985)
47 J. Fünfschilling, D.F. Williams: Photochem. Photobiol. 26, 109 (1977)
48 J. Hala, I. Pelant, L. Parma, K. Vacek: J. Lumin. 24/25, 803 (1981)
49 R. Avarmaa, A. Suisalu: Eesti NSV Tead. Akad. Toim. Füüs. Matem. (Proc. Estonian SSR Acad. Sci.) 33, 333 (1984) (in Russian)
50 R.A. Avarmaa, A.P. Suisalu: Opt. Spektrosk. 56, 53 (1984) [English transl.: Opt. Spectrosc. USSR 56, 32 (1984)]
51 R.A. Avarmaa, K.K. Rebane: Spectrochim. Acta 41A, 1365 (1985)
52 K.K. Rebane, R.A. Avarmaa: Chem. Phys. 68, 191 (1982)
53 I. Renge, K. Mauring, P. Sarv, R. Avarmaa: J. Phys. Chem. 90, 6611 (1986)
54 R.A. Avarmaa, K.H. Mauring: Opt. Spektrosk. 41, 670 (1976) [English transl.: Opt. Spectrosc. USSR 41, 393 (1984)]
55 R. Avarmaa, K. Mauring, A. Suisalu: Chem. Phys. Lett. 77, 88 (1981)
56 R. Jaaniso: Eesti NSV Tead. Akad. Toim. Füüs. Matem. (Proc. Estonian SSR Acad. Sci.) 31, 161 (1982) (in Russian)
57 K. Mauring, R. Avarmaa: Chem. Phys. Lett. 81, 446 (1981)
58 K.N. Solov'ev, E.I. Zalesskii, V.N. Kotlo, S.F. Shkirman: Pis'ma Zh. Eksp. Teor. Fiz. 17, 463 (1973) [English transl.: JETP Lett. 17, 332 (1973)]
59 E.I. Zen'kevich, A.M. Shul'ga, A.V. Chernook, G.P. Gurinovich, E.I. Sagun: Zh. Prikl. Spektrosk. 42, 772 (1985)
60 S.S. Dvornikov, V.A. Kuz'mitskii, V.N. Knyukshto, K.N. Solov'ev, A.E. Turkova, E.G. Ivanov: Khim. Fiz. 4, 889 (1985)
61 K.K. Rebane, R.A. Avarmaa, K.H. Mauring, R.V. Jaaniso: in Sovremennye aspekty tonkostrukturnoj i selektivnoj spektroskopii (MGPI imeni V.I. Lenina, Moskva 1984) p. 63

62 R.V. Jaaniso, R.A. Avarmaa: Zh. Prikl. Spektrosk. 44, 601 (1986)
63 K. Rebane, R. Avarmaa: in Molecular Spectroscopy of Dense Phases, Proc. 12th European Congress of Molec. Spectrosc. (Elsevier, Amsterdam 1976) p. 459
64 V.G. Maslov, A.S. Chunayev, V.V. Tugarinov: Biofiz. 25, 925 (1980)
65 V.G. Maslov, A.S. Chunayev: Molek. Biol. 16, 604 (1982)
66 J. Friedrich, H. Scheer, B. Zikendraht-Wendelstadt, D. Haarer: J. Am. Chem. Soc. 103, 1030 (1981)
67 P.J. Angiolillo, L.S. Leigh, J.M. Vanderkooi: Photochem. Photobiol. 36, 133 (1982)
68 R. Avarmaa, I. Renge, K. Mauring: FEBS Lett. 167, 186 (1984)
69 I. Renge, K. Mauring, R. Avarmaa: Biochim. Biophys. Acta 766, 501 (1984)
70 O.B. Belyaeva, E.R. Personova, F.F. Litvin: Photosynthesis Research 4, 81 (1983)
71 K. Mauring, I. Renge, R. Avarmaa: FEBS Lett. (1987) (in print)
72 U. Even, J. Magen, J. Jortner, J. Friedman, H. Levanon: J. Chem. Phys. 77, 4374 (1982)
73 A.Yu. Borisov, A.M. Freiberg, V.I. Godik, K.K. Rebane, K.E. Timpmann: Biochim. Biophys. Acta 807, 221 (1985)
74 I.Renge, K. Mauring, R. Avarmaa: J. Lumin. 37, 207 (1987)

Spectral Hole Burning (SHB): Scientific and Practical Applications

J. Kikas

Institute of Physics, Estonian SSR Acad. Sci.
SU-202400 Tartu, Estonian SSR, USSR

Abstract

A classification of experiments on zero-phonon SHB is given and some related methodical problems are discussed. SHB applications in fluorescence-line-narrowing spectroscopy to improve its selectivity and the possibilities of controlling the SHB process are considered. Some technical applications of SHB in the field of transformation and recording of optical signals are discussed.

1. Introduction

The potentialities of zero-phonon-line (ZPL) spectroscopy [1] are best revealed by the spectral hole burning [2,3] based on the pioneering studies [4,5].

The physical essence of SHB consists in a spectrally highly selective phototransformation of impurity molecules in low-temperature solid matrices absorbing monochromatic radiation via their narrow ZPL. Stable (long-lived) narrow holes in the inhomogeneous distribution function (IDF) of zero-phonon transition energy are, in fact, (negative) images of homogeneous ZPL. They allow one to study a large variety of subtle phenomena in solids, otherwise obscured by the inhomogeneous broadening of spectra. SHB ensures, in principle, the maximum elimination of inhomogeneous broadening - its spectral resolution is determined by the homogeneous linewidth of the purely electronic line (PEL) and at $T \to 0$ it reaches up to 10^{-4} - 10^{-5} cm^{-1} for allowed dipole transitions. This enables one, for example, to use the Doppler scanning of the light wavelength in experiments on SHB, which is an unprecedented method in solid state optical spectroscopy [6]. An essential reason of great and ever growing popularity of SHB is a large variety of its micromechanisms, ranging from irreversible photodissociation of impurity molecules [7] to the finest structural rearrangements in the vicinity of the impurity [5,8].

Scientific applications of SHB lie, first of all, in the field of solid state spectroscopy of very high spectral resolution. SHB has proved to be particularly versatile in the studies of glassy media characterized by a large (100 - 1000 cm^{-1}) inhomogeneous broadening of impurity spectra.

The most active investigations have been carried out on the thermal broadening of ZPL. One of the basic results of the ZPL theory, the approaching of the radiative limit by PEL homogeneous width at $T \to 0$, has experimentally been verified by SHB [7,9].

On the other hand, peculiarities of the PEL thermal broadening in amorphous impurity solids [2,10,11,12] have been stimulating a notable theoretical activity in this field (see [13]) and creating novel theoretical conceptions for describing such materials (two-level systems, fractal structures and fractions). To most recent achievements belong the observation of a non-elementary temperature dependence of PEL homogeneous width at ultralow (down to 50 mK) temperatures [14] and the discovery of systems preserving narrow (of the order of 10^{-3} cm^{-1}) holes after an intermediate heating up to room temperatures [15].

SHB has successfully been applied in the studies of biomolecules and molecular, including photosynthesizing, systems in vivo (see [16,17]).

Along with scientific applications, more and more attention is being paid to possible technical applications of SHB in the field of optical information storage and transformation of optical signals. SHB opens a new - frequency (spectral) - dimension for information storage. Novel functional possibilities of the latter are most impressively revealed by the time-and-space-domain holography [18,19].

2. Spectroscopy of Zero-Phonon Holes

Narrow zero-phonon holes can be used as sensitive indicators of different processes and interactions in solids. An outline of experiments on spectroscopy of no-phonon holes is given in Table 1. In what follows, some related methodological problems are considered.

Spectral holes are characterized by three main parameters carrying various physical information: the maximum frequency ν_o, width δ and area S (not considering the polarizational characteristics, see, e.g. [20]). We are not going to deal here with the burning process itself (see [2]) and we suppose that during the recording stage the number of burnt-out impurities is conserved. Possible changes of the hole area are then related to the changes of the Debye-Waller factor and will not be considered here. The main spectroscopic information is then carried by the hole frequency and width.

Two principle types of experiments are used in the spectroscopy of zero-phonon holes (Figure 1). In the first case (1), the hole is burnt in and recorded under the

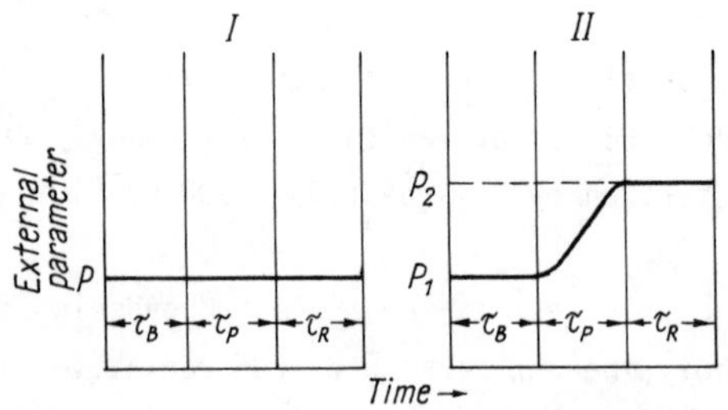

Figure 1: Schematic diagrams of the experiments on determining the dependence of the homogeneous width (I) and spectral shift (II) of ZPL on the external parameter P by SHB. τ_B - burning time, τ_R - recording time, τ_P - time interval between burning and recording

Table 1. Experiments on zero-phonon SHB

Parameter varied (for a fixed impurity)	Experimental scheme (according to Figure 1)	
	I	II
	Parameter measured	
	Holewidth	Shift (dispersion of shifts)
1. Matrix	[10,21,22,23,24]	[25]
2. Transition	[23,26]	[27,28,29]
3. Position within the inhomogeneous band	[14,26]	
4. Temperature	[10,14,21,22,30, 31,32,33]	[31]
5. Electric field		[32,34,35,36]
6. Magnetic field		[37,38,39]
7. Uniaxial stress		[40,41]
8. Hydrostatic pressure		[41]
9. Time ($\tau_B + \tau_P + \tau_R$)		[42,43]

same (fixed) physical conditions. In this case the hole frequency coincides with that of the burning monochromatic light ν_L and the whole spectroscopic information is contained in the holewidth δ. Under certain limitations for the intensity and the dose of the burning light the holewidth coincides with the double homogeneous linewidth of ZPL, 2Γ, [20]. In such experiments, the dependences $\Gamma(P)$ of the homogeneous linewidth on various parameters P, characterizing physical conditions, can be determined. The relation of the homogeneous linewidth Γ to the characteristic times of longitudinal (T_1) and transversal (T_2') relaxation is simple:

$$\Gamma = (2\pi T_1)^{-1} + (\pi T_2')^{-1} \quad . \tag{1}$$

This enables one to study the influence of various factors on the processes of phase and energetic relaxation.

In experiments of the second type (Figure 1, II), the hole is burnt and recorded at different physical conditions (at values P_1 and P_2 of an actual parameter P). In this case the hole frequency will not coincide with that of the burning light and the shifts $\Delta\nu = \nu_0 - \nu_L$ contain additional physical information. The holewidth is given by

$$\delta = \Gamma(P_1) + \Gamma(P_2) \tag{2}$$

in case of a uniform spectral shift of all impurities. Particularly, this category includes the experiments on recording the non-resonant (to the burning light) holes related to some other zero-phonon transitions. If, for example, ν_L is within the 0-0 absorption band, and holes are recorded within the vibronic absorption bands, then the difference $\nu_0 - \nu_L$ gives us vibronic frequencies in the excited electronic state [27,28].

The situation is more complicated in the case of media, where even the molecules with a fixed zero-phonon transition frequency reveal an essential dispersion of spectral shifts. This is, particularly, the case of the media with isotropic orientational distribution of impurities affected by external vector fields (electric field [34,35], uniaxial stress field [40]). Therefore, only the mean value of spectral shifts is obtainable strightforwardly from such experiments. The dispersion of spectral shifts gives rise to an additional broadening of holes ($\delta > \Gamma(P_1) + \Gamma(P_2)$). The magnitude of such dispersion can be determined by comparing the results of experiments I, II (see, for example, the determination of an inhomogeneous dispersion of vibrational frequencies [27]).

It must be noted that a reversible (cyclic) variation of macroconditions does not necessarily result in a reversible change of microconditions. In the spectroscopy of zero-phonon holes this manifests itself as a broadening of holes after a cyclic change of physical conditions (e.g. tempertaure [25]): $P_1 \rightarrow P_2 \rightarrow P_1$. In Table 1, this is indicated as a variation of the matrix in experiments II.

If during the measurement cycle slow (as compared to the lifetime of the excited electronic state) processes of structural rearrangement take place in the matrix, resulting in spectral diffusion, the total time of the measurement cycle ($\tau_B + \tau_P + \tau_R$) becomes also an essential parameter [42,43].

3. SHB and Site Selection by the Fluorescence Line Narrowing

The fluorescence line narrowing by monochromatic laser excitation [44,45] (see also [17]) as a predecessor of SHB, lacks one of the most essential advantages of the latter: the persistence of the achieved selection. The molecules selected by monochromatic excitation become different from the remaining ones only during their stay in the excited state. After relaxation to the ground state they become indistinguishable within the whole ensemble of impurity molecules. Only by fixing the selection by a chemical reaction (or some other transformation of the impurity centre), i.e. by SHB, the lifetime of the "spectral memory" can be essentially prolonged.

Moreover, SHB enables one to further improve the very method of fluorescence line narrowing to increase its selectivity. It is well known [46] that even a monochromatic excitation within the purely electronic absorption band will not ensure a comlete (with an accuracy of Γ) spectral selection of impurities: a comparable amount of molecules is excited via their broad phonon sidebands (PSB), which also contribute to the observable fluorescence spectrum. However, by incorporating SHB, such contribution can be eliminated [47,48]. The essence of the method proposed in [47] consists in the following. In the presence of hole burning processes a notable redistribution of the relative intensities of the line and sideband takes place in the monochromatically-excited fluorescence spectra in the course of irradiation [10] (Figure 2). This is due to a far more effective burning of the impurities excited via ZPL. By choosing an appropriate dose of irradiation a situation can be achieved,

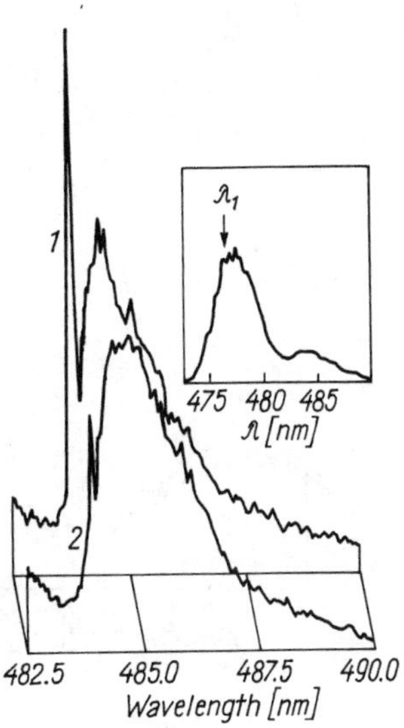

Figure 2: Fluorescence spectra of ethanol:tetracene solid solution (T = 4.2 K) excited by a 476.5 nm-line of an Ar^+-laser before (1) and after (2) a 1000 s laser irradiation. Insert - the inhomogeneous fluorescence spectrum of the same system

when the number of impurities excited via ZPL is essentially decreased already, while the impurities excited via PSB are still practically unburnt. The key point here is the large ratio of peak absorption coefficients in homogeneous ZPL and PSB, reaching $10^4 - 10^5$ for PEL. In the difference spectrum, obtained by subtracting the fluorescence spectrum measured after burning from the initial one (measured before burning), only the contribution from the impurities excited via ZPL is preserved. Thus, such selection by absorption coefficients leads also to the increase of spectral selectivity. The method described can also be applied to selectively-recorded excitation spectra: in this case an intermediate burning has to be carried out at the frequency of selective (narrow-band) registration.

By using these methods one can obtain spectra of an impurity subensemble with a fixed (with an accuracy of Γ) purely electronic transition frequency. According to the simplest treatment of the inhomogeneous broadening (shifts of homogeneous spectra without any shape distortion) [49] such spectra are considered as homogeneous ones. However, it follows from rather general considerations that even within such subensemble impurities can differ essentially by other spectral parameters [50]. On the example of a solid solution of Zn-tetraphenylporphin in an ether-alcohol glass it was demonstrated [51] that SHB can be used for further selection within a subensemble of molecules with a fixed purely electronic transition frequency. By a mono-

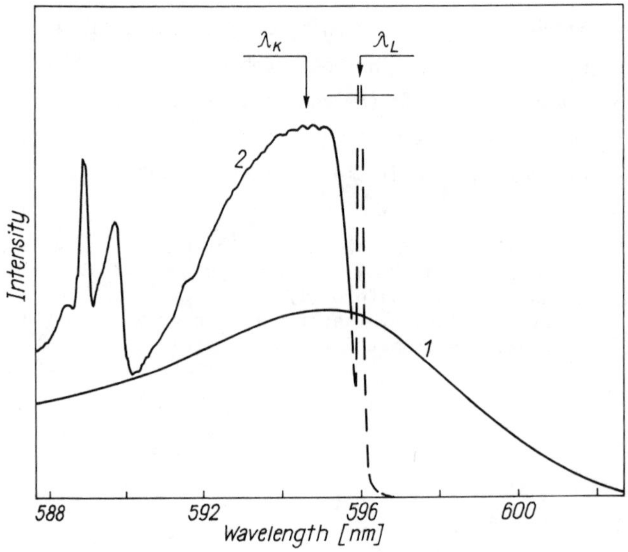

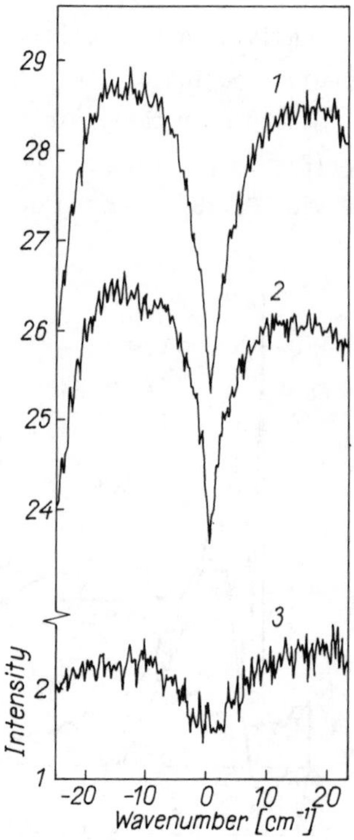

Figure 3: Excitation spectra of the solid solution of Zn-tetraphenylporphin in a diethyl ether-butanol mixture (5:2 volume ratio, T = 5 K) under broadband (1) and selective (2) recording of fluorescence. Wavelengths of the selective recording (λ_L = 595.96 nm) and burning in the sideband (λ_K = 594.48 nm) are indicated

Figure 4: Holes burnt in the sideband in the excitation spectrum at λ_K recorded before (1) and after (2) burning at the wavelength of selective recording λ_L and their difference (3). The last spectrum represents the hole in the "homogeneous" sideband, revealing its hidden inhomogeneous structure

chromatic excitation it was possible to burn comparatively broad (≈10 cm^{-1}) holes in the structureless sideband in the "homogeneous" excitation spectrum obtained by the difference method described above (Figures 3 and 4). This gives evidence about the existence of a narrow peculiarity in the spectrum of the impurity molecule, not correlated with the PEL and contributing to the broad sideband. Its physical nature may be related to the splitting of the degenerated excited electronic state into two electronic components, the higher of which is broadened due to a fast relaxation.

It should be added that in the cases considered above SHB manifests itself as a specific method of non-linear spectroscopy, more exactly, of saturational spectroscopy. Close methods have been developed also on the basis of other saturation mechanisms (optical saturation of $S_1 \leftrightarrow S_0$ transition [52], triplet bottleneck saturation [53]). For these cases, however, higher excitation intensities are needed.

4. Control of the SHB Process

For many scientific and technical applications of SHB it is highly desirable to disconnect the steps of selective excitation and phototransformation and to have a possibility of an independent control (gating) of the latter. In principle, all effects capable of accelerating or inhibiting molecular transformations, can be used. Most possible factors (temperature, external fields), however, will yield also an undesirable broadening and shift of holes (see Chap. 2). Probably the only acceptable solution is the two-step phototransformation. In this case, the first absorbed light quantum transfers the molecule spectrally selectively into a chemically nonactive excited state (e.g. S_1), and only by an absorption of the second quantum the impurity molecule is transferred into a chemically active higher state (directly from S_1 or after an intermediate relaxation into the lowest triplet state). In the case of different wavelengths of the quanta used (two-colour light-gated hole burning [54]), the recording of holes by using only the first-step excitation is possible, which does not result in any hole distortion due to the continuing of SHB process.

For the first time the two-step nature of the SHB process was established for a solid solution of tetracene in the isopropylalcohol-ether glass [10], where a quad-

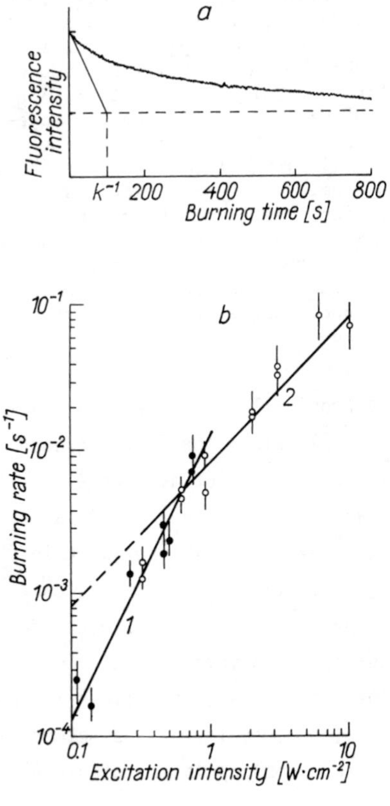

Figure 5: a. Kinetics of ZPL intensity in the fluorescence spectrum of ethanol:tetracene solid (glassy) solution on laser excitation ($\lambda_L = 476.5$ nm, $I = 0.7\ W \cdot cm^{-2}$). Dashed horizontal line - asymptotic value of the fluorescence intensity at large burning times; the determination of the burning rate K is indicated; b. 1 - The dependence of the burning rate on excitation intensity for the above-mentioned system, 2 - the same for n-hexane:tetracene solid (polycrystalline) solution

ratic dependence on the excitation intensity (of 476.5 nm Ar^+-laser line) was observed for the decay rate of the vibronic ZPL line intensity (Figure 5). In this experiment, the wavelengths of the first and second steps of excitation coincided, which does not enable one to infer the possibility of a two-colour SHB in this system. Two-colour light-gated SHB was recently demonstrated for $BaCl:Sm^{2+}$ [15] and $LiGa_5O_8:Co^{2+}$ [55] impurity crystals, boric acid:carbazole [54] and polymethyl-methacrylate:tetramesosubstituted Zn-tetrabenzporphin [56,57] solid solutions.

SHB is based on spectrally selectively induced phototransformations of impurity molecules. The solution of the inverse problem of spectrally selective inhibition of an impurity phototransformation is of both scientific and practical interest as well. It has been demonstrated that for this purpose an inverted scheme of two-step two-colour SHB can be used [58]. The first quantum is used to populate non-selectively the photochemically active excited electronic state, and the second quantum with a lower energy stimulates transitions to photochemically non-active vibronic levels of the ground electronic state (Figures 6, 7). The method of inverted SHB enables one, in

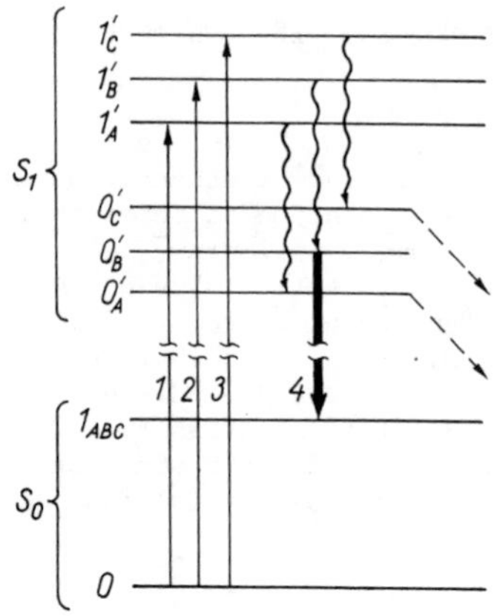

Figure 6: The scheme of actual energetic levels and transitions in experiments on inverted SHB by light-stimulated depopulation of S_1-state. The system studied is polystyrene:octhaethylporphin solid solution at 5 K. Energetic levels for three molecules (A,B,C) from the spectrally inhomogeneous ensemble are depicted. Straight solid arrows indicate the transitions induced by non-selective excitation ($\lambda_L = 585$ nm, $\delta\lambda_L = 7$ nm, $I = 2\cdot10^8$ $W\cdot cm^{-2}$) within the inhomogeneous vibronic band of $S_1 \leftarrow S_0$ transition (1, 2, 3) and selectively light-stimulated ($\lambda_L = 650.33$ nm, $\delta\lambda_L = 5\cdot10^{-2}$ nm, $I = 10^8$ $W\cdot cm^{-2}$) transition (4) to a vibrational level of the ground electronic state (S_0). Wavy arrows - vibrational relaxation, dashed arrows - impurity phototransformation. Some relaxation channels (spontaneous fluorescence, singlet-triplet conversion) are not indicated

principle, to determine the homogeneous widths of vibronic fluorescence lines by the widths of antiholes and, hence, the rates of vibrational relaxation in the ground electronic state. It should be noted that such information is not available in site-selected fluorescence spectra because of vibrational inhomogeneity [27]. An alternative spectral method is offered by IR SHB [8,24], however, it will not work in the case of chemically non-active vibrations in the ground state.

If ground-state vibrations are active in the SHB process one should expect the formation of holes instead of antiholes in the experiments on inverted SHB. In general, the method enables one to study the activity of ground state vibrations in the process of impurity phototransformations. The spectral selectivity of the method excludes the influence of possible by-processes (e.g. two-step population of higher electronic states).

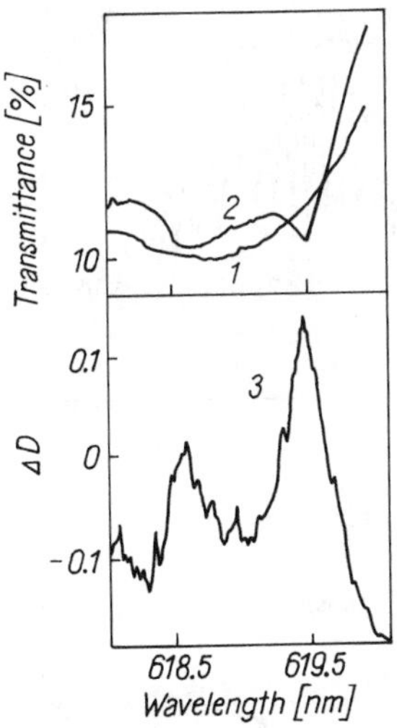

Figure 7: Transmission spectra of the polystyrene:octaethylporphin solid solution before (1) and after (2) irradiation with two lasers, as indicated in Figure 6, 3 - the difference absorption spectrum

The inverted SHB is worth mentioning as a method for narrowing the inhomogeneous distribution function as well. Such narrowing may be especially effective in the systems with a large number of mutually phototransformable (quasi-)equilibrium impurity configurations in the ground state. Inverted SHB yields a persistent increase of the spectral density of impurities.

5. Technical Applications of SHB

5.1. Narrow-Band Optical Filters

A narrow photobleached band in the inhomogeneous absorption spectrum serves, at a sufficient optical density of the sample, as an effective narrow-band transmission filter. As to the transmission bandwidth ($\geq 10^{-3}$ cm^{-1}), such filters are superior to the best interference filters. Owing to the absorption principle of operation, SHB filters have one more advantage as compared to interference filters: they can operate in noncollimated beams. Their transmission profiles are rather insensitive to the incident angle of the light. By using SHB, a prompt tuning of the transmission wavelength is available, and filters of complex transmission profiles can be manufactured easily. In Figure 8, the spectral characteristics for a particular realization of such a filter are depicted.

5.2. Binary Information Recording

SHB opens new prospects for optical information storage. Diffraction makes it impossible to focus a light beam on areas less than of the order of λ^2, λ being the wavelength of the light. This limits the surface density of information for all kinds of optical storage. For spectrally nonselective binary recording (a certain spot irra-

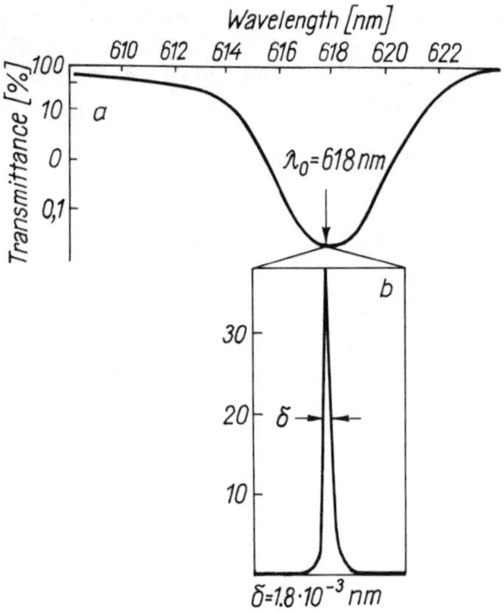

Figure 8: The transmission spectrum of a sample of the polystyrene:octaethylporphin solid solution (a) and the transmission band of a narrow-band filter produced by SHB (b)

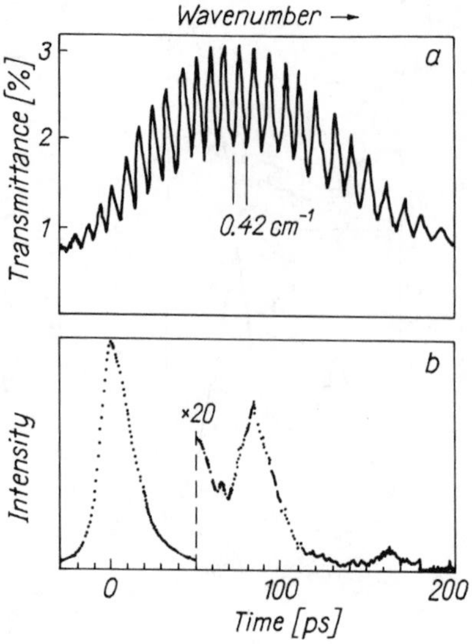

Figure 9: A grating of spectral holes in the transmission spectrum of the polystyrene:free-base tetra-(tret.-butyl)-porphirazine solid solution (a) and the temporal replay of the sample to a pulse excitation (b)

diated/not irradiated) at $\lambda = 500$ nm one obtains the limit value of the surface density $4 \cdot 10^8$ bit/cm^2. SHB offers an additional - spectral - dimension for information storage: a simple hole in the inhomogeneous spectrum can carry one bit of information [59]. The increase of the recording (surface) density is then determined by the number of (independent) holes, which one can burn in the inhomogeneous spectrum. The most optimistic estimates for the surface density of spectrally selective recording reach the values of 10^{14} bit/cm^2, which exceed the respective values for spectrally nonselective recording by 6 orders of magnitude. The burning of 1600 holes in the inhomogeneous $S_1 \leftarrow S_0$ absorption band of octaethylporphin molecules in a polystyrene matrix has been demonstrated experimentally [60]. To burn a grating of spectral holes an excitation with a sequence of mutually coherent picosecond pulses was used, while the contrast of the grating was determined by the relative intensity of the time response to pulse excitation [61] (Figure 9, see also [62]).

5.3. Analog Information Recording

Historically the above-considered method for spectrally selective binary recording was the first proposal to use SHB in optical information storage. However, the lack of threshold (both by intensity and irradiation dose) for burning processes in most

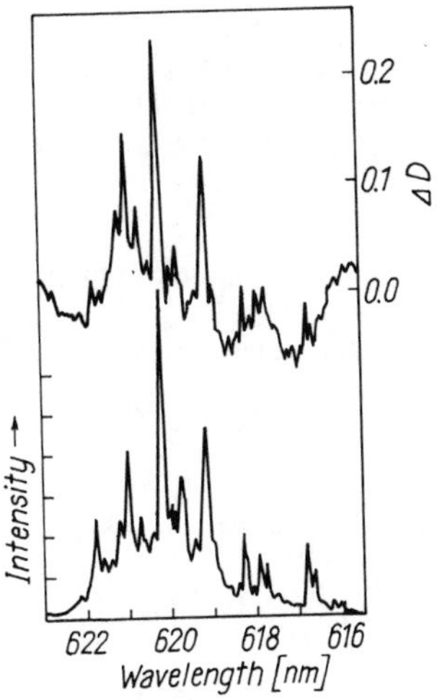

Figure 10: The fluorescence spectrum of the n-hexane:octaethylporphin solid solution (T = 5 K) recorded by DFS-24 spectrometer (below) and by using a spectrally selective photoactive medium (above)

of the materials studied makes them more suitable for an analog recording. Such possibilities have to the most extent been realized by the method of four-dimensional (spatial-temporal) holography on the basis of SHB [18,19].

SHB can be used in other cases as well, where a high-resolution recording of the spectral composition of a light field is needed. In Figure 10, the fluorescence spectrum of the solid solution of octaethylporphin in n-hexane (Shpol'skii system) is presented, recorded by both SHB and a common spectrograph method [63].

6. Conclusion

The trend of scientific research based on SHB, has by now gone far beyond the scope of traditional "contemplative" spectroscopy. In fact, a new field of research has appeared at the borderland of optics, solid state physics and photochemistry, which deals with the production of materials with given optical properties by higly-selective action of light on solids.

Many problems of the spectroscopy of zero-phonon lines, earlier considered as purely academical ones (formation and temperature dependence of ZPL and PSB, inhomogeneous broadening), are now of major technical and technological importance. This will stimulate, without doubt, further progress in the fundamental research on ZPL spectroscopy.

References

1 K.K. Rebane: Elementarnaya teoriya kolebatel'noj struktury spektrov primesnykh tsentrov kristallov (Nauka, Moskva 1968) [English transl.: Impurity Spectra of Solids (Plenum Press, New York 1970)]
2 L. Rebane, A. Gorokhovskii, J. Kikas: Appl. Phys. B29, 235 (1982)

3 J. Friedrich, D. Haarer: Angew. Chem. Int. Ed. Engl. 23, 113 (1984)
4 A.A. Gorokhovskii, R.K. Kaarli, L.A.Rebane: JETP Lett. 20, 474 (1974)
5 B.M. Kharlamov, R.I. Personov, L.A. Bykovskaya: Opt. Commun. 12, 191 (1974)
6 K.K. Rebane, V.V. Palm: Opt. Spektrosk. 57, 381 (1984) [English transl.: Opt. Spectrosc. USSR 57, 229 (1984)]
7 H.de Vries, D.A. Wiersma: Chem. Phys. Lett. 51, 565 (1977)
8 M. Dubs, Hs.H. Günthard: Chem. Phys. Lett. 64, 105 (1979)
9 A.A. Gorokhovskii, L.A. Rebane: Izv. Akad. Nauk SSSR Ser. Fiz. 44, 859 (1980) [English transl.: Bull. Acad. Sci. USSR Phys. Ser. 44, 148 (1980)]
10 A.A. Gorokhovskii, J.V. Kikas, V.V. Palm, L.A. Rebane: Fiz. Tverd. Tela 23, 1040 (1981) [English transl.: Sov. Phys. - Solid State 23, 602 (1981)]
11 H.P.H. Thijssen, R.van den Berg, S. Völker: Chem. Phys. Lett. 97, 295 (1983)
12 A.A.Gorokhovskii: in this issue
13 M.A. Krivoglaz: in this issue
14 A.A. Gorokhovskii, V.H. Korrovits, V.V. Palm, M.A. Trummal: Pis'ma Zh. Eksp. Teor. Fiz. 42, 249 (1985) [English transl.: JETP Lett. 42, 307 (1985)]
15 A. Winnacker, R.M. Shelby, R.M. Macfarlane: Opt. Lett. 10, 350 (1985)
16 R.A. Avarmaa, K.K. Rebane: Spectrochim. Acta 41A, 1365 (1985)
17 R. Avarmaa: in this issue
18 A.K. Rebane, R.K. Kaarli: Izv. Akad. Nauk SSSR Ser. Fiz. 48, 545 (1984) [English transl.: Bull. Acad. Sci. USSR Phys. Ser. 48, 128 (1984)]
19 P. Saari, R. Kaarli, A. Rebane: J. Opt. Soc. Am. B3, 527 (1986)
20 M.D. Levenson, R.M. Macfarlane, R.M. Shelby: Phys. Rev. B22, 4915 (1980)
21 H.P.H. Thijssen, R.E.van den Berg, S. Völker: Chem. Phys. Lett. 103, 23 (1983)
22 S. Voelker: J. Lumin. 36, 251 (1987)
23 A.I.M. Dicker, S. Völker: Chem. Phys. Lett. 87, 481 (1982)
24 W.E. Moerner, A.J. Sievers, A.R. Chraplyvy: Phys. Rev. Lett. 47, 1082 (1981)
25 J. Friedrich, D. Haarer, R. Silbey: Chem. Phys. Lett. 95, 119 (1983)
26 S. Voelker, R.M. Macfarlane: Chem. Phys. Lett. 61, 421 (1979)
27 A.A. Gorokhovskii, J. Kikas: Opt. Commun. 21, 272 (1977)
28 B.M. Kharlamov, L.A. Bykovskaya, R.I. Personov: Chem. Phys. Lett. 50, 407 (1977)
29 J. Friedrich, D. Haarer: J. Chem. Phys. 79, 1612 (1983)
30 S. Voelker, R.M. Macfarlane, A.Z. Genack, H.P. Trommsdorf, J.H.van der Waals: J. Chem. Phys. 67, 1759 (1977)
31 S. Voelker, R.M. Macfarlane, J.H.van der Waals: Chem. Phys. Lett. 53, 8 (1978)
32 W.E. Moerner, A.R. Chraplyvy, A.J. Sievers, R.H. Silsbee: Phys. Rev. B28, 7244 (1983)
33 A.A. Gorokhovskii, L.A. Rebane: Opt. Commun. 20, 144 (1977)
34 A.P. Marchetti, M. Scozzafawa, R.H. Young: Chem. Phys. Lett. 51, 424 (1977)
35 V.D. Samojlenko, N.V. Razumova, R.I. Personov: Opt. Spektrosk. 52, 580 (1982) [English transl.: Opt. Spectrosc. USSR 52, 346 (1982)]
36 R.T, Harley, R.M. Macfarlane: J. Phys. C16, 1507 (1983)
37 A.I.M. Dicker, J. Dobkowski, M. Noort, S. Völker, J.H.van der Waals: Chem. Phys. Lett. 88, 135 (1982)
38 A.I.M. Dicker, M. Noort, H.P.H. Thijssen, S. Völker, J.H.van der Waals: Chem. Phys. Lett. 78, 212 (1981)
39 A.I.M. Dicker, M. Noort, S. Voelker, J.H.van der Waals: Chem. Phys. Lett. 73, 1 (1980)
40 W. Richter, G. Schulte, D. Haarer: Opt. Commun. 51, 412 (1984)
41 T. Sesselmann, W. Richter, D. Haarer: J. Lumin. 36, 263 (1987)
42 A.A. Gorokhovskii, V.V. Palm: Pis'ma Zh. Eksp. Teor. Fiz. 37, 201 (1983) [English transl.: JETP Lett. 37, 237 (1983)]
43 W. Breinl, J. Friedrich, D. Haarer: J. Chem. Phys. 80, 3496 (1984)
44 A. Szabo: Phys. Rev. Lett. 25, 924 (1970)
45 R.I. Personov, E.I. Al'shitz, L.A. Bykovskaya: Pis'ma Zh. Eksp. Teor. Fiz. 15, 609 (1972) [English transl.: JETP Lett. 15, 431 (1972)]
46 J. Kikas: Chem. Phys. Lett. 57, 511 (1978)
47 R. Jaaniso: Eesti NSV Tead. Akad. Toim. Füüs. Matem. (Proc. Estonian SSR Acad. Sci.) 34, 277 (1985) (in Russian)
48 J. Fünfschilling, D. Glatz, I. Zschokke-Gränacher: J. Lumin. 36, 85 (1986)
49 R. Avarmaa: Eesti NSV Tead. Akad. Toim. Füüs. Matem. (Proc. Estonian SSR Acad. Sci.) 23, 238 (1974) (in Russian)

50 T.B. Tamm, P.M. Saari: Chem. Phys. Lett. 30, 219 (1975)
51 J.V. Kikas, R.V. Jaaniso: Fiz. Tverd. Tela 26, 1861 (1984) [English transl.: Sov. Phys. - Solid State 26, 1128 (1984)]
52 J.V. Kikas, A.B. Treshchalov: Chem. Phys. Lett. 98, 295 (1983)
53 R. Jaaniso, J. Kikas: Chem. Phys. Lett. 123, 169 (1986)
54 H.W.H. Lee, M. Gehrtz, E.E. Marinero, W.E. Moerner: Chem. Phys. Lett. 118, 611 (1985)
55 R.M. Macfarlane, J.-C. Vidal: Phys. Rev. B34, 1 (1986)
56 O.N. Korotaev, E.I. Donskoj, V.I. Glyadkovskii: Opt. Spektrosk. 59, 492 (1985) [English transl.: Opt. Spectrosc. USSR 59, 298 (1985)]
57 T.P. Carter, C. Bräuchle, V.Y. Lee, M. Manavi, W.E. Moerner: J. Phys. Chem. 91, 3998 (1987)
58 J. Kikas, I. Sildos: Chem. Phys. Lett. 114, 44 (1985)
59 G. Castro, D. Haarer, R.M. Macfarlane, H.P. Trommsdorff: US Patent N 4101976
60 J.V. Kikas, R.K. Kaarli, A.K. Rebane: Opt. Spektrosk. 56, 387 (1984) [English transl.: Opt. Spectrosc. USSR 56, 238 (1984)]
61 P. Saari: in this issue
62 A.K. Rebane, R.K. Kaarli, P.M. Saari: Pis'ma Zh. Eksp. Teor. Fiz. 38, 320 (1983) [English transl.: JETP Lett. 38, 383 (1983)]
63 A.A. Gorokhovskii, J.V. Kikas, R.V. Jaaniso: to be published

Broadening of Zero-Phonon Lines of Impurities in Glasses

A.A. Gorokhovskii

Institute of Physics, Estonian SSR Acad. Sci.
SU-202400 Tartu, Estonian SSR, USSR

Abstract

Methods, models, and experimental results on the homogeneous broadening of zero-phonon lines in the spectra of impurity molecules in organic glass-like matrices at low and superlow temperatures are reviewed.

1. Introduction

Thanks to their extraordinary narrowness, high Q-factor and peak intensity, zero-phonon lines (ZPL) serve as sensitive indicators of impurity states in solids at low temperatures [1]. This general statement is vividly conformed by ZPL of the impurities introduced into disordered glass-like matrices. In this case a ZPL behaviour is revealed [2 - 4], which is fundamentally different from that in ordered crystals and which reflects the basic pecularities in the structure of glasses.

The following will be a brief survey of the methods, theoretical models and experimental results existing up to now for the low-temperature behaviour of the homogeneous broadening of the narrowest - purely electronic-ZPL in the impurity spectra of organic and, to a less extent, inorganic glasses. The results obtained in the department of the spectroscopy of molecules and crystals at the Institute of Physics of the Estonian SSR Acad. Sci., will be discussed in more detail. Note that some recent surveys on the broadening of ZPL in glasses have been published in [5].

2. Low-Temperature Properties of Glasses and Anomalous Behaviour of ZPL

In contemporary science, a glass-like state is defined as a nonequilibrium metastable state of a liquid with a frozen structure. In contrast to ideal crystals, glasses are characterized by the absence of the long-range order. Nevertheless, it seems that at liquid helium temperatures, when actual phonon wavelengths are so large that the structure of the matter can be ignored and a solid considered homogeneous, the thermal behaviour of the heat capacity and the thermal conductivity of glasses and crystals must be very close [6]. However, in 1971, it was discovered [7] that in glasses in the temperature region of 0.1 - 1 K linear law was valid for thermal capacity ($C \sim E$) and quadratic, for thermal conductivity ($Q \sim T^2$), contrary to the one predicted in the Debye model and observed for crystals $C,Q \sim T^3$. Subsequently, a number of experiments were carried out (see reviews in [8]), confirming the results of [7]. A number of theoretical models were proposed to explain these facts. The model proposed in [9,10] became most popular and was best confirmed experimentally. In this model, groups of atoms or other structural elements were assumed to be present in glasses, which are able to perform tunnel transitions between two (or possibly more) nearly equivalent equilibrium positions separated by an energy barrier. As a result of tunnelling, the levels of vibrational motion near the equilibrium position are split and the spectrum of such system consists of a set of doublets. At low temperatures only two levels of the ground vibrational state are important and this is why such tunnelling system is often called a two-level system (TLS). The energetic splitting of a TLS depends on the parameters of the double-well potential, which change randomly in glasses. This causes an extensive distribution of TLS with respect to energies, which in models [9,10] is assumed to be uniform ($N(E) = const$) up to the lowest energies. Note that a single TLS can evidently be well modelled by molecular impurities able to perform reorientational motions in solids [11].

Such anomalous (in comparison with crystals) low-temperature properties of glasses have attracted the attention of spectroscopists. The difficulties due to large ($\sim$100 - 1000 cm^{-1}) inhomogeneous broadening of impurity spectra in glasses (as a result of their structural disorder) were overcome with the help of the methods of selective spectroscopy - the luminescence line narrowing (LLN) [12,13] and spectral hole burning (SHB) [14,15]. The vibronic spectra of impurities, obtained by these methods, were in basic details close to those predicted by the theory [1] and observed in experiments on crystals: an intensive and narrow ZPL was accompanied by a broad phonon wing. Anomalies were detected in finer though fundamental details, such as homogeneous ZPL widths, Γ, and their temperature dependences. These characteristics are determined by the phase relaxation time T_2 of the excited electronic state of the impurity, as well as by relaxation mechanisms, as follows: $\Gamma = (\pi T_2)^{-1} = (2\pi T_1)^{-1} + (\pi T_2')^{-1}$, where T_1 is the energy (longitudinal) relaxation time and T_2', the "pure" dephasing time [16]. This anomalous behaviour, that seems to have been reported for the first time for inorganic systems in [2] and for organic ones in [3, 4], and later in a number of other works [17-41], consists in the following:

(i) At liquid helium temperatures Γ exceeds essentially (by 1 - 3 orders of magnitude) the values of the homogeneous linewidths observed for the same impurities in crystalline matrices [3,30] and approach the limiting value $\Gamma_0 = (2\pi T_1)^{-1}$ determined for optical transitions only by the lifetime T_1 of the electronic state [1,42];

(ii) At temperatures considerably lower than the Debye temperature weak temperature dependences of the homogeneous ZPL width, $\Gamma(T) \sim T^n$, with $0.5 \leq n \leq 2.6$, are observed (in some cases a broadening law is observed which changes in 0.1 - 1 K range [30,38]), contrary to the $\sim T^7$ predicted by the adiabatic theory for crystals in the case of interaction with acoustical phonons [1,42-48], or to

$\exp(-\hbar\Omega/\kappa T)$ in the case of interaction with a pseudolocal vibration with the frequency $\Omega > \kappa T/\hbar$ [45-47], both observed in experiment.

As an example, Figure 1 [4] depicts the spectra of a hole and the temperature dependences $\Gamma(T)$ for organic impurities in the matrices of three types - glass-like

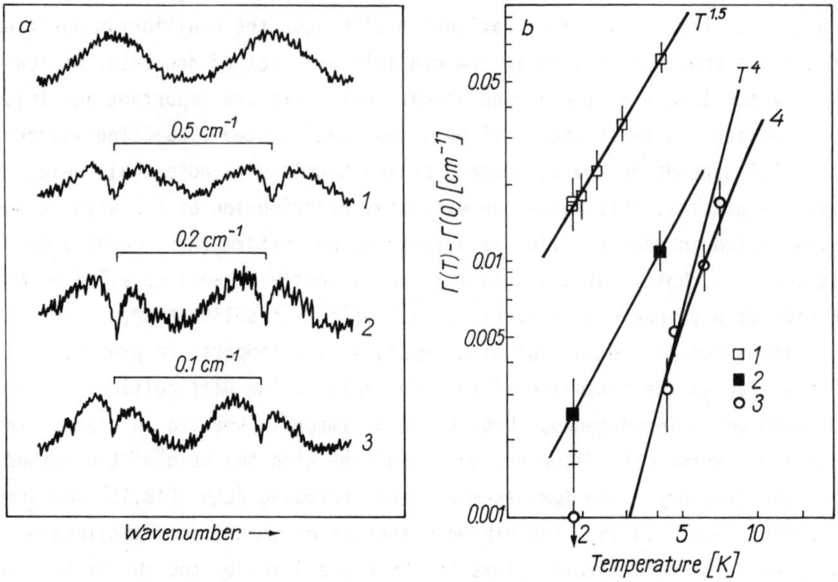

Figure 1: Holes in the excitation spectrum at 1.7 K (a) and temperature dependences of the homogeneous linewidth (b) of H_2-tetra-4-tert.-butylphthalocyanine in different matrices: 1 - alcohol glass; 2 - n-nonane; 3 - tetradecane. Curve 4 is calculated by using $\Gamma(T) - \Gamma(0) = 0.098 \cdot n(\Omega)[n(\Omega) + 1]$, $n(\Omega) = [\exp(h\Omega/kT)-1]^{-1}$ with quasi-local mode frequency $\Omega = 10\ cm^{-1}$. The upper left spectrum is the initial one before burning

(inhomogeneous width $\Delta \approx 170\ cm^{-1}$), poor crystalline ($\Delta \approx 150\ cm^{-1}$) and good crystalline (Shpol'skii system with $\Delta \approx 5\ cm^{-1}$). It can be seen that the matrix structure has a strong influence on the homogeneous widths and temperature dependences of a ZPL.

3. Photoburning of Persistent Holes in the Spectra of Glasses

The main progress in the investigation of the mechanisms of ZPL broadening in organic glasses and polymers is connected with the development of the SHB method (see [49-57] and the review by J.Kikas in this issue). In case of SHB, monochromatic radiation excites a part of centres in the inhomogeneously-broadened band with ZPL frequencies distributed in conformity with the homogeneous absorption spectrum. As a result of the final probability for the excited centres to go over to a state with a changed ZPL frequency (to go over to another electronic state, to undergo a photochemical transformation, etc.), in the inhomogeneous distribution of ZPL frequencies a hole is formed. The lifetime of the hole is determined by that of the photoproduct and its duration at low temperatures and in the absence of illumination may really be long - for hours, years, and even centuries [51,55]. The hole can be recorded in the spectra of absorption luminescence. The shape of the obtained hole may be rather complicated [51,58]. However, owing to a high peak intensity of ZPL, at the excitation frequency a narrow peculiarity - a "zero-phonon" hole - can be discriminated in it. This hole corresponds to the absence of the centres absorbing the pumping radiation through their ZPL. In a simple model, where the influence of inhomogeneity is reduced only to a shift of ZPL frequencies, the spectrum of such hole in absorption is described by the convolution $I_h \sim F_h * F_k$ [51,58], where F_h is the shape of a narrow hole in the function of inhomogeneous distribution of centres over ZPL frequencies, which depends on the homogeneous ZPL absorption spectrum F_k as well as on the irradiation dose and intensity [51,59]. In view of a long lifetime of the hole, the times of the burning and recording of the hole spectrum may considerably exceed the lifetime of the electronic state T_1. This allows the burning and recording processes to be separated, ensuring an unquestionable methodological preference of SHB as compared to the LLN on recording homogeneous ZPL spectra: the difficulties in connection with the separation of luminescence from the scattered laser light on resonance measurements are removed (see below). On the other hand, a new considerable time appears - the measurement time τ_m including the times of the burning, delay and recording of the hole, which usually makes $\tau_m \simeq 10^{-2} - 10^3$ s. If during τ_m the matrix structure remains unchanged (that, evidently, holds for crystals in a thermodynamically-equilibrium stable state), then, on burning by small doses of not very intensive light [59], $F_h \sim F_k$ and, consequently, $I_h \sim F_k * F_K$, i.e. the spectrum of the hole carries information about the homogeneous ZPL spectrum. In case of a Lorentzian shape of a ZPL, the shape of the hole I_h is also Lorentzian, while its width $\delta_h = 2\Gamma_k$, where Γ_k is the homogeneous width of the ZPL in the absorption.

In case of glasses, the structure of which is thermodynamically metastable, during the time τ_m, the processes of the relaxation of the structure to a new more stable state are possible. In the course of such structural relaxation, the surroundings of impurity centres undergo random changes, which brings about also random changes of ZPL frequencies. By analogy with the random changes of the Larmor frequencies of spins in the theory of electron spin echo this phenomenon is called spectral diffusion [60,61]. It leads to the change of the hole shape in time - its broadening and becoming more shallow [3,61-63]. At sufficiently high temperatures, near the vitrification point, the contribution to relaxational processes is made by the motion of various structural groups of glass (molecular groups, separate chain segments, etc.), each of which is characterized by its own relaxation time τ_r. With the lowering of temperature thermal motions are gradually frozen. At liquid helium temperatures the main role in relaxation processes is played by tunnelling motions, including the motion of the matrix elements which form TLS.

In accordance with models [9,10] TLS display a wide variety of the relaxation times τ_r. According to [64], an estimate of the minimal relaxation time at T = 1 K gives for quartz $\tau_{r,min} \simeq 10^{-9}$ s, for polystyrene, $\tau_{r,min} \approx 10^{-12}$ s. A comparison with the optical lifetime of the allowed transition, $T_1 \simeq 10^{-8}$ s, indicates that the transitions in TLS can perform a fast ($\tau_r < T_1$) as well as slow ($\tau_r > T_1$) random modulation of the ZPL frequency. In case of fast modulation the spectral diffusion contributes to the homogeneous ZPL width. The slow motions lead to spectral diffusion that adds some quantity $\delta_{sp.diff.}$ to the inhomogeneous broadening of the hole.

Thus, in the case of glasses F_h and, consequently, the hole spectrum I_h, carry information about the fast processes of the homogeneous ZPL broadening as well as about the slow processes of spectral diffusion, broadening the hole inhomogeneously as the holewidth is $\delta_h = 2\Gamma + \delta_{sp.diff.}$.

In connection with what was said above there arises a problem of separating the contributions by the homogeneous broadening and by slow spectral diffusion in the hole spectrum. It becomes possible by measuring the homogeneous ZPL widths by independent fast ($\tau_m \leq T_1$) methods and by comparing the results with those of SHB [19,35, 36,62,65] (see Sects. 5.3 and 5.5). One of such methods is LLN [12,13,66] (see also review papers [52,67]). The line in the luminescence spectrum on monochromatic excitation is described by the convolution $I_f \sim F_k * F_f$ [58] and it contains information about the homogeneous spectra of luminescence, F_f, and about the absorption lines through which the excitation of luminescence occurs. Besides, the inhomogeneous distribution of the frequencies of intramolecular vibrations brings about an additional broadening of luminescence linewidths [68]. It follows that to obtain a homogeneous spectrum of a purely electronic line (PEL) it is necessary to measure the luminescence spectrum in resonance with the excitation frequency. Then we have to detect the line on a very strong background of scattered laser excitation. Naturally, in the case of excited states of long lifetimes we can chop the excitation and perform measurements in the luminescence afterglow as it has been done for strongly-forbidden transitions of rare-earth ions [2,17,21] and for singlet-triplet transitions of organic molecules [32,33,62,69]. SHB meets well the requirements for the studies of PEL provided the measurement times τ_m are short enough to eliminate the spectral diffusion. When τ_m is shortened down to the electronic state's lifetime T_1, we actually have to deal with the (transient) spectral holes of optical saturation [70]. Another possibility is the time-domain methods - photon echo, free induction decay and their analogs - which work the better the narrower are the homogeneous linewidths. These methods are entirely founded on non-linear effects, i.e. a strong field of light is needed and it may additionally distort the picture, especially in the case of subtle experimental situations.

To conclude, an essential circumstance for glasses as strongly-disordered systems must be emphasized. In the methods of SHB and LLN, the selection of certain-type centres from the whole inhomogeneous ensemble takes place only by ZPL frequencies. However, a certain group of centres with fixed ZPL frequencies may have inhomogeneous distribution with respect to other ZPL parameters, e.g. with respect to the width or shape of the homogeneous spectrum or integral intensity. In this case, the simple relations of the convolution, written above for the observed "homogeneous" spectra, do not hold any more and the observed hole or luminescence line is $I_h^o \sim \langle F_k * F_k \rangle$ or $I_f^o \sim \langle F_k * F_f \rangle$, respectively. Averaging is performed over the inhomogeneous parameter of the ZPL spectrum [51,53,71] or, in a more general case [71], over the arrangements of the atoms of the medium, which determine the shape of the homogeneous spectra, F_k and F_f. This results in complicated forms of the observed spectra and in the violation of simple relations of the type $\delta_h = 2\Gamma_k$. For instance, in the simplest case of a Lorentzian shape of F_k with a uniform distribution over the widths between Γ_1 and Γ_2, the resulting spectrum I_h^o is not Lorentzian and its width $\delta_h^o = 2\sqrt{\Gamma_1\Gamma_2}$ [71]. An opposite situation is also possible: the Lorentzian shape of the observed spectrum, I_h^o, is obtained by an appropriate averaging over the parameters of the TLS of non-Lorentzian homogeneous spectra [72,73].

4. Mechanisms of the Homogeneous Broadening of ZPL in Glasses

The role of TLS in the formation of the optical spectra of glasses was indicated already in 1976 [2]. By now a number of models has been worked out, taking into account the interaction of the impurity with TLS and the vibrations of the sceleton of glass. The results are listed in Table 1.

Theoretical models differ by the consideration of various interactions on calculating the spectrum in one centre and by the methods of the subsequent averaging over the inhomogeneous parameters of the matrix. It is supposed that the impurity and TLS may interact via electrical or elastic forces of various multiplicity. The vibrations of the matrix interact with the impurity through phonon-induced transitions in TLS. The direct interaction of the impurity with the impurity-induced quasilocal vibration (QV) [72,76,78] as well as with the intrinsic QV of the disordered matrix [72] are also taken into account. In [79], the interaction of TLS with fractons - excitations of the fractal structure of the disordered matrix of polymers and glasses [85] - is considered. On calculating the linewidths the averaging of the spectral widths of single centres is performed over various parameters of neighbouring TLS and QV: $\bar{\Gamma} = \langle\Gamma\rangle$ [35,74,75,77-80]. *Krivoglaz* [72] has criticized this approach, maintaining that not the linewidths but distributions of various centres should be averaged and that in the SHB (LLN) method the ZPL should be determined as the widths δ_h^o (δ_f^o) of the resulting spectral distribution $\langle I_h \rangle$ ($\langle I_f \rangle$). The main contribution into the broadening is made by the shifts of the electronic levels of the impurity on fluctuational transitions in TLS. Most essential are the TLS transitions bringing about a modulation slow with respect to the dephasing rate $T_2 = 2/\pi\delta_o$ ($T_2 < \tau_r < T_1$). The shape of the ZPL under consideration depends on the nature of the impurity-TLS

Table 1. Temperature dependence of the homogeneous linewidths $\delta(T)$ of ZPL in glasses according to various versions of the theory by the n-th power-law approximation, $\delta(T) \sim T^n$

Counterparts in interaction	Temperatures (K)							References
	$T < T_c$			$T > T_c$			T_c	
	Fields of interactions							
	d-d	d-q	q-q	d-d	d-q	q-q		
TLS + AP	2			1			T_D	[74]
TLS + AP	2			1			E_{max}	[75]
Impurity QV				2			Ω	[76]
TLS + AP	1	7/4	11/5	0	1/4	2/5	E_{max}	[77]
TLS + AP, QV	2			1-1.5			E_{max}	[78]
TLS + fractons	1	4/3	23/15					[79]
TLS + AP	2.5 - 4			0.5-1			0.01-0.001 T_D	[80]
TLS + OP	$e^{-\Omega_{min}/T}$			1			1/2 Ω_{min}	[80]
TLS + AP (Redfield's theory)	1	7/4	11/5					[35]
Impurity QV	$e^{-\Omega_{min}/T}$			2 ($\tau_r \ll \delta^{-1}$) 1 ($\tau_r \gg \delta^{-1}$)			1/2 Ω_{min}	[72]
Intrinsic QV	2	3 ($T > 0.1\ T_D$)	4	1	4/3	5/3	1/2 Ω_o	[72]
TLS + AP	1	4/3	5/3					[72]
TLS + AP, sp. diff.	1							[81]
TLS + AP, sp. diff.	2 ($\tau_r > T_1$)							[82]
TLS + AP sp. diff.	1.18 (Direct pr.) 1.3 (Raman pr.)							[83]
TLS + AP sp. diff.	1	4/3	5/3					[73]
Crystal with a two-well potential	1-2							[84]

Notations: T_c - critical temperature distinguishing the low- and high-temperature regimes of broadening; d - dipole; q - quadrupole; AP - acoustic phonons; OP - optical phonons; Ω - frequency of OP or QV modes; T_D - the Debye temperature; E - energy splitting of TLS; Ω_o comes from the equation $\tau_r(\Omega_o) = \delta^{-1}$; the values of n have been calculated with the assumption that the density of states $N(E) = const$; if $N(E) \sim E^{\mu}$, then $\delta(T) \sim T^{n+\mu}$; fundamental constants $\hbar = 1$, $k = 1$

interaction: in the case of a dipole-dipole interaction it is Lorentzian, in the case of higher multiplicity the line becomes narrower in the central part and decays slower on the wings [72].

The results of [72] are in agreement with those of [73,81] which are based on the models of spectral diffusion, i.e. random changes of ZPL frequencies, induced by fluctuational transitions in TLS as a result of their interaction with phonons. In these theories, the shape of ZPL is time-dependent: the line broadens with time.

Osad'ko [84] has considered a model of glass as a crystal, in which the impurity and the atoms of the matrix are sited in a two-well adiabatic potential. The "distortion" of phonons on tunnelling is taken into account. The interaction of the optical electron with the tunnelling degrees of freedom leads to a complementary phonon contribution into broadening, which results in complicated dependences between T and T^2 with one or two cross-overs. The predicted lineshape is Lorentzian. The theory [84] is concerned, in principle, with one impurity centre and assumes the role of averaging to be negligible.

Thus, theoretical models predict a broad spectrum of possible temperature dependences of the ZPL linewidth on T^0 up to T^4 and $\exp(-\hbar\Omega/kT)$. Since the density of elementary excitations of the matrix depends on temperature, the study of $\delta(T)$ can provide information on mechanisms and relative contributions of actual interactions. The lineshape and its time dependence are important additional information sources. It should be noted that the theories are directly applicable to the lineshapes measured via LLN. When interpreting the experimental data of SHB the slow spectral diffusion processes should additionally be taken into account.

5. Experimental

5.1. Inorganics

Data on homogeneous linewidths of ZPL in the spectra of rare-earth ions in inorganic glasses are presented in Table 2 (see also review paper [86]). In the pioneering paper [2], the $^5D_0 - {}^7F_0$ transition of Eu^{3+} in silicate glass and in the single crystal $YAlO_3$ has been studied. The linewidth in the case of glass was by two orders of magnitude broader than in the crystal and depended on temperature as $T^{1.8\pm0.2}$, while in the crystal the dependence was exponential in accordance with the Orbach mechanism. In [19], the first persistent spectral holes in inorganics were studied. The accumulated photon echo experiments on Pr^{3+} in silicate glass indicated that there was no spectral diffusion in the time interval $10^{-4} - 10^{2}$ s. Nd^{3+} impurities in silica optical fiber were studied at subkelvin temperatures 0.1 - 1 K by the photon echo method [20]. The $T_2^{-1} \sim T^{1.3}$ dependence found in [20] has in [73] been interpreted as the result of elastic dipole-dipole interaction of the impurity with TLS, the latter having the density of states $N(E) \sim E^{0.3}$. That interpretation was supported by three-pulse photon echo experiments [87]. The measured linewidths of ZPL in inorganics are rather small (at 0.1 K in [20] $(\pi T_2)^{-1} = 0.2$ MHz) but, because of the for-

Table 2. Temperature dependence of the ZPL linewidths $\delta(T)$ in inorganic glasses

Glasses	Temperature (K)	n in approximation $\delta(T) \sim T^n$	Method	Reference
Eu^{3+} in silicate	7 - 80	1.8 ± 0.2	LLN	[2]
Pr^{3+} in BeF_2 and GeO_2	7 - 300	1.9 ± 0.2	LLN	[17]
Nd^{3+} in silicate	10 - 70	2.2	Accumulated echo	[18]
Eu^{3+} in silicate	1.6	The point occurs on the dependence $T^{1.8}$ [2]	SHB	[19]
Pr^{3+} in silicate	1.6 - 20 4 - 20	1.0 ± 0.2 1	SHB Accumulated echo	[19]
Nd^{3+} in silicate	0.1 - 1	1.3	Two-pulse echo	[20]
Yb^{3+} in silicate and phosphate	6.5 - 40 40 - 70	1.3 ± 0.1 1.9 ± 0.1	LLN	[21]

bidden character of the transitions under study, they are still much broader than the limit values determined by the energy relaxation time of the excited electronic state.

5.2. Organics

Indications to large holewidths in disordered organic matrices were contained already in the first publications on SHB in glasses [15,88,89], performed by a relatively low resolution. The reasons of how such values came about were not discussed in these papers. A comparison of the holewidths [3] and their temperature dependences [4] (Figure 1) for the same molecule, introduced into a glassy and a crystalline matrix, indicated the essential role of the structural disorder in organic systems.

Subsequently a number of molecules in different glassy and polymeric organic hosts have been studied at low and ultralow temperatures. The data are collected in Table 3. The curves of the temperature dependence of ZPL widths, if interpreted via power laws, vary from $T^{0.5}$ to $T^{2.6}$. Most of the studies have been performed for metal-free porphyrins as impurity molecules. In this case the hole creation mechanism is a 90^0 photo-induced reorientation of a proton pair in the core of the molecule [90] and the quantum yield is rather high - 10^{-2} - 10^{-3}. The holewidths in the spectra of these molecules in different matrices at 4.2 K are 0.007 - 0.2 cm^{-1}.

The systems indicated by asterisks in Table 3 are characterized by broader holes (1 - 5 cm^{-1}) and lower quantum yields of hole creation (below 10^{-4}). The latter feature requires higher intensities of photoburning light (10^{-3} - 10^{-2} $W \cdot cm^{-2}$ in comparison with 10^{-6} - 10^{-5} $W \cdot cm^{-2}$ for metal-free porphyrins). The mechanisms of triplet

Table 3. Linewidths δ and their temperature dependence δ(T) in organic glasses and polymers

Material	Temperature (K)	n in approximation $\delta(T) \sim T^n$	Method	δ, cm^{-1} (T = 4.2 K)	Reference
1	2	3	4	5	6
H_2-tetra-4-tert.-butylphthalocyanine - alcohol glass	1.7-4.2	1.5±0.3	SHB	0.12	[4]
Tetracen - alcohol and glycerol glasses*	2-14	1	SHB	3-6	[22]
1,4-dihydroxyanthraquinone - alcohol and boric acid glasses*	2-25	2	SHB	0.6-2	[23]
Cresylviolet - polyvinyl alcohol*	1.8-11	1.1±0.1	SHB	1	[24]
H_2-phthalocyanine - polystyrene	2-25	1.8±0.3	SHB	0.1	[25]
H_2-phthalocyanine - polyethylene	2-16	2.6±0.2	SHB	0.1	[25]
H_2-chlorin - polyvinylbutanol	0.4-6	1.6±0.2	SHB	0.04	[26]
Tetracen - anthracene glass*	2.5-15	1.9±0.1	SHB	5	[27]
Tetracen - 9,10 diphenylanthracene glass*	2.5-10	1.4±0.1	SHB	6	[28]
H_2-porphin, H_2-chlorin, dimethyl-S-tetrazine, resorufin, cresylviolet in several glasses and polymers	0.4-18	1.3±0.1	SHB	0.1-0.2	[29]
H_2-porphin - polyethylene	0.3-2.2 2.2-4.2	1.03±0.05 1.3±0.1	SHB	0.007	[30]
H_2-porphin - diglycerol	0.3-1.1 1.1-4.2	1.0±0.1 1.3±0.1	SHB	0.027	[30]
Resorufin - ethanol glass	0.4-4.2	1.3±0.1	SHB	0.1 (phase I) 0.01 (phase II)	[31]
Pyren - 1-bromo-butane	1.7-7	1.8±0.2	LLN (in phosphorescence)	0.04	[32]
Coronen and 5-bromoacenaphthene - 1-bromo-butane	1.8-4.2	1.25±0.20	LLN (in phosphorescence)	0.06-0.3	[33]
H_2-phthalocyanine - sulfuric acid glass	1.2-10	1.8±0.1	SHB	0.07	[34]
Pentacene - polymethylmethacrylate*	1.5-12 1.5-12	1.30±0.05 2	Accumulated echo SHB	0.04 1	[35]

Table 3. (continued)

1	2	3	4	5	6
Resorufin - ethanol glass	1.5-11.4	Combination of exp($-\hbar\Omega/kT$) and T, ($\hbar\Omega$ = 19 cm^{-1})	Two-pulse echo	0.035	[36]
H_2-chlorin - amorphous silica	1.7-5.7	1.45±0.07	SHB	0.05	[37]
Oxazine-4 - amorphous silica	1.7-5.7	1.27±0.05	SHB	0.09	[37]
H_2-octaethylporphin - polystyrene (for λ_1 and λ_2)	0.05-0.1 0.1-0.2 0.2-1.5	2.4±0.5 2.6±0.5 0.5±0.1 0.75±0.10 1.07±0.05 1.22±0.05	SHB	0.03 (T = 1.5 K)	[38-40]
H_2-octaethylporphin - polystyrene	4-16 5-11	1.8±0.1 3.0±0.1	SHB LLN	0.4 0.2 (T = 7 K)	[41]
H_2-octaethylporphin - polymethylmethacrylate	0.04-0.1 0.1-2	2.3±0.5 1.20±0.05	SHB	0.016 (T = 1.9 K)	[40]
H_2-monoazaethioporphyrin IV - polystyrene	0.1-2	1.12±0.05	SHB	0.045 (T = 1.9 K)	[40]

saturation and (or) photo-induced rearrangements of the matrix in the vicinity of the impurity are probably involved. For these systems it is rather difficult to consider the holewidths to be close to the homogeneous linewidths of ZPL.

In [29], the "universal law" $\delta_h \sim T^{1.3}$ has been established for a considerable number of systems. For some of them $T^{1.3}$ dependence goes over to $\delta_h \sim T$ for the temperature interval between T = 2 - 1 K up to T = 0.3 K. At the latter temperature the holewidth for H_2-porphin in semicrystalline polymer polyethylene is already quite close to the limit value determined by the energy relaxation time T_1 [30]. An analysis of the experimental curves on the basis of theoretical models [78,79,83] is given in review paper [57]. Some correlation between the holewidth and the size of the molecular sidegroup linked to the main chain of the polymeric molecule of the matrix has been established: the holes are narrower in the matrices with smaller sidegroups.

A pentacene molecule in a polymethylmethacrylate matrix has been studied by photon echo and SHB [35]. The photon echo measurements gave $T_2^{-1} \sim T^{1.3}$, while the result via SHB was $\delta_h \sim T^2$ with holewidths 2 - 3 times broader than $(\pi T_2)^{-1}$. The photon echo data match the model of dipole-dipole interaction between impurity and TLS with the density of states $N_{TLS}(E) \sim E^{0.3}$. The SHB shows an additional broadening due to slow spectral diffusion. An analogous conclusion has been drawn from a comparison of the results on photon echo and SHB for resorufin in ethanol glass [36].

In conclusion note that the homogeneous ZPL in organic materials are broader than those in inorganic systems. It is probably due to two circumstances: the density of low-frequency vibrational models is high (because of massive molecules bound by relatively weak intermolecular forces) and the softness of the structure of the matrix provides plenty of possibilities for low-energy rearrangements (including the quasi-rotational and tunnelling ones).

5.3. Homogeneous Linewidth of the $T_1 - S_0$ Transition of Pyrene on Glassy and Crystalline Hosts

The essential role of spectral diffusion was demonstrated on the example of a pyrene molecule introduced into glass-like 1-bromo-butanol (BrBu) matrix [32,62]. The three-fold-degenerate T_1-state of the molecule is split into three sublevels and quite a complicated structure of the hole and resonance fluorescence spectra, depending on the spin-lattice relaxation times, becomes possible [62,69]. But the relatively small probability (≤ 0.1) of transitions to two of the three spin sublevels and the supposed slow spin-lattice relaxation rate make a good approximation for interpreting the spectra as corresponding to one single transition. The holewidth at 1.7 K (Figure 2a) was $\delta_h = 0.07$ cm^{-1}, while the ZPL in phosphorescence was tenfold narrower, $\delta_{ph} = 0.007$ cm^{-1} (Figure 3a). We are inclined to interpret the result $\delta_h >> \delta_{ph}$ as the indication of spectral diffusion going on even at liquid helium temperatures fast enough to shift the ZPL position in $\tau_m = 10^3$ s and slow enough to influence to considerably smaller extent the excited triplet state during its lifetime $T_1 = 80$ ms. This conclusion corresponds to the experimental evidence that the hole measured at 4.2 K after 30 min in dark (Figure 2c, d) has considerably gained in width and lost in depth. Slow processes of spectral diffusion on the time scale $10^4 - 10^5$ s have also been detected in the systems of a chinozarin molecule in alcohol glasses and others [55,63].

The temperature dependence of the homogeneous linewidth calculated as $\Gamma = \delta_{ph}/2 - \delta_{ex}$ is shown in Figure 3b. The curve may be approximated as $\Gamma(T) \sim T^{1.8 \pm 0.2}$. This dependence close to a quadratic one may be interpreted (see Table 1) as a dipole-dipole interaction with low-frequency ($\Omega \leq 1$ cm^{-1}) quasilocal modes of the impurity [72,76]. Another possibility is the quadrupole-quadrupole interaction between the impurity

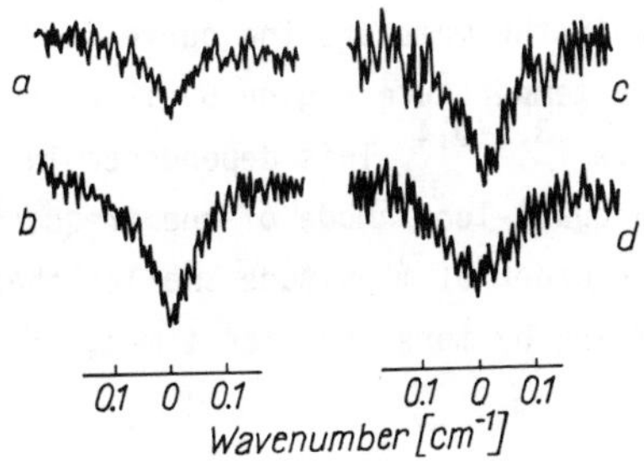

Figure 2: Holes in the phosphorescence excitation spectrum of Py-BrBu: after burning for 10 (a) and 30 min (b) at 1.8 K, immediately after burning (c) and after keeping in dark for 30 min (d) at 4.2 K

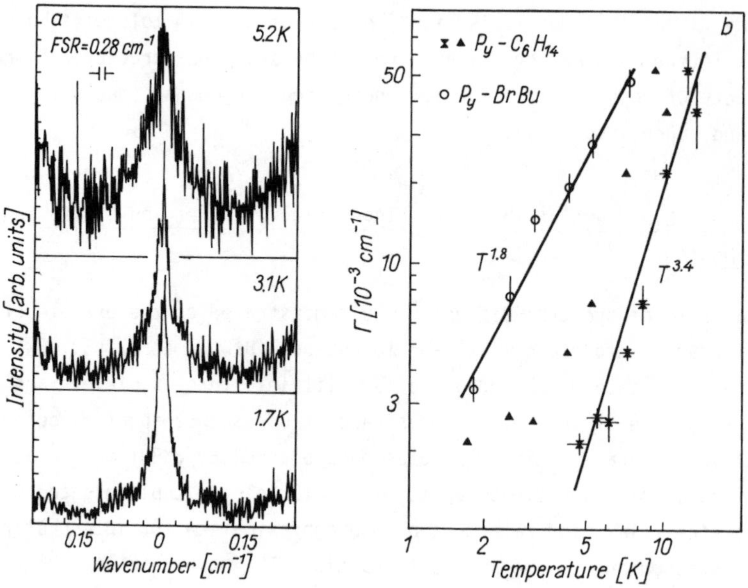

Figure 3: a. 0-0 phosphorescence lines of Py-BrBu under resonance laser excitation, measured in the afterglow by means of a scanning Fabry-Perot interferometer; b. temperature dependences of the homogeneous linewidth of 0-0 phosphorescence line for pyrene in glassy (circles) and crystalline matrices (measured by using external (▲) and untrinsic (⧗) thermometers)

and TLS [72]. The choice between these two possibilities depends on the lineshape. It is difficult to get this choice from the phosphorescence spectrum because we have to expect, especially at the wings of the ZPL, some contribution from weak transitions to other sublevels. At 1.7 K the homogeneous linewidth $\Gamma = 0.0035 \pm 0.0005$ cm^{-1}. As far as the superfine interaction is weak and the spin-lattice, triplet-singlet and photochemical radiationless relaxations are slow, the linewidth is governed mainly by dephasing processes.

For comparison in Figure 3b the dependence $\Gamma(T)$ for pyrene in polycrystalline matrix of n-hexane (Shpol'skii system) is presented [91]. In this system, an overheating of about 3 K as a result of laser excitation ($\sim$1.5 W/mm^2) is observed. The real temperature of the impurity was determined according to the temperature dependence of spin-lattice relaxation rates as an intrinsic thermometer. The curve is quite different from the one for BrBu matrix. In the temperature region 5 - 12 K, the linewidth increases rapidly and follows the power law $T^{3.4\pm0.4}$. This dependence indicates the modulation mechanism of broadening by a quasi-local mode of the frequency $\Omega \simeq 20$ cm^{-1}. At $T = 5$ K, $\Gamma = 0.002$ cm^{-1} and it is by an order of magnitude smaller than for BrBu. As that value exceeds the superfine splitting by more than ten times, it is evidently determined by pure dephasing.

5.4. Photoburning of Holes in the Impurity Spectra of Amorphous Polymers at 0.04 - 2 K

The best-known specific features of the disordered glass-like structures - peculiarities in heat capacity and heat transport - become prominent below 1 K. It is reasonable to think that the same holds for optical properties and the subkelvin temperatures are of special interest also in the studies of ZPL dephasing. The first measurements of the broadening of photoburnt holes down to 0.05 K were performed in the spectra of H_2-octaethylporphin (OEP) impurity in polystyrene (PS) (Figure 4a), i.e. in an amorphous polymer as a matrix [38,39]. In what follows, improved data on OEP-PS as well as new data on OEP in polymethylmethacrylate (PMMA) and H_2-monoazaethioporphyrin (MAEP) in PS will be discussed [40].

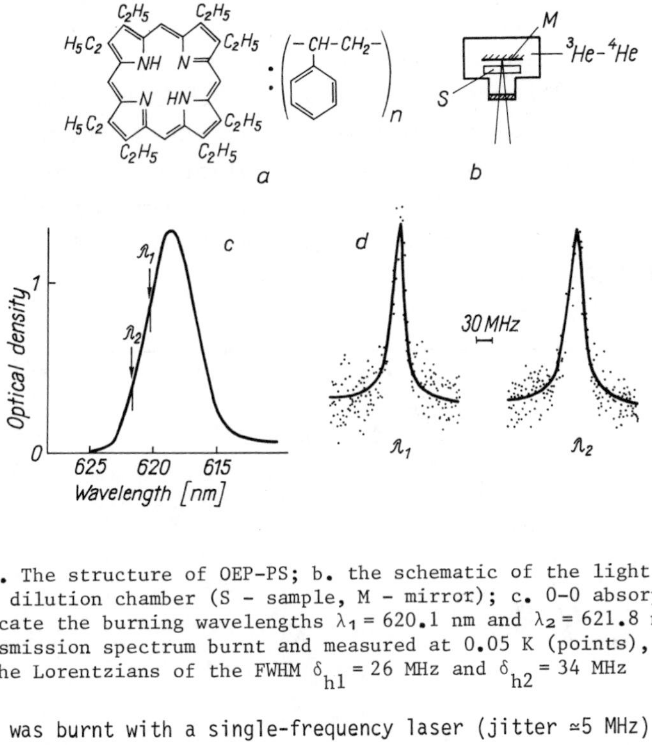

Figure 4: a. The structure of OEP-PS; b. the schematic of the light beam passing through the dilution chamber (S - sample, M - mirror); c. 0-0 absorption band (the arrows indicate the burning wavelengths $\lambda_1 = 620.1$ nm and $\lambda_2 = 621.8$ nm); d. the holes in the transmission spectrum burnt and measured at 0.05 K (points), the solid lines represent the Lorentzians of the FWHM $\delta_{h1} = 26$ MHz and $\delta_{h2} = 34$ MHz

The hole was burnt with a single-frequency laser (jitter ≃5 MHz) and registered with a multichannel analyzer in the transmission spectrum by scanning the laser over the burning region and by accumulating the signal. To attain temperatures 0.04 - 2 K an optical $^3He/^4He$ dilution refrigerator was used. The laser beam was led in vertically, passed through the object which was placed into the mixing chamber (Figure 4b), reflected back from the prism, passed through the object for the second time and then led out from the cryostat into a photomultiplier. The experimental set-up has been described in more detail in [39].

The holes for OEP-PS were burnt in the long-wavelength side of the 0-0 absorption band (λ_{max} = 618.5 nm, $\Delta \simeq 150$ cm^{-1}) at λ_1 = 620.1 nm and λ_2 = 621.8 nm (Figure 4c, d). The holewidths at T = 0.05 K were $\delta_1 = 26.0 \pm 1.5$ MHz and $\delta_2 = 34.0 \pm 1.5$ MHz for λ_1 and λ_2, respectively, that considerably exceeds the limiting holewidth, determined by the lifetime of the S_1 state, $T_1 = 17.5 \pm 1.0$ ns: $\delta_o = 2\Gamma_o = (\pi T_1)^{-1} = 18$ MHz (in the results published earlier [38,39], an inaccurate value, δ_o = 10.8 MHz, was used). T_1 depends weakly on temperature in the range of interest. Therefore, the lifetime-limited part in the holewidth can be excluded by analysing the difference $\delta(T) - \delta_o$. These dependences are presented on a double logarithmic scale in Figure 5a. The broadening follows a complicated law with two cross-over points at T ≃ 0.1 and T ≃ 0.2 K, slightly different for λ_1 and λ_2. The powers are given for the approximation of separate parts to the power law $\delta(T) - \delta_o = T^n$. In the temperature region T < 0.1 K, n = 2.4 and 2.6 (±0.5), at T = 0.1 - 0.2 K n = 0.75 and 0.5 (±0.1), at T > 0.2 K n = 1.07 and 1.22 (±0.05) for λ_2 and λ_1, respectively. Extrapolation of $\delta(T)$ to T → 0 indicates that the holewidths approach the limiting value only at T = 0.01 - 0.03 K.

In the case of OEP-PMMA, the holes were burnt at λ_1 = 618.5 nm and λ_2 = 616 nm. The obtained widths and their temperature dependences coincide within the limits of the measurement accuracy. At T = 0.04 K δ = 16 - 18 MHz, which is close to the lifetime-limited width δ_o = 15 MHz (τ_1 = 21 ns). The temperature dependences of $\delta(T) - \delta_o$ (Figure 5b) exhibit one cross-over point at T = 0.1 K, where the broadening law changes from n = 2.3 ± 0.5 to n = 1.20 ± 0.05.

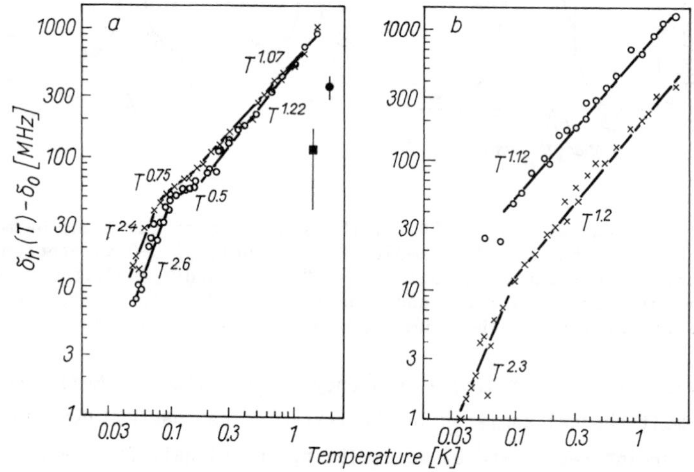

Figure 5: Temperature dependences of $\delta_h(T) - \delta_o$ (δ_o - lifetime-limited holewidths) in the spectra of OEP-PS for λ_1 = 620.1 nm (o) and λ_2 = 621.8 nm (x) (a) and of MAEP-PS (o) and OEP-PMMA (x) (b). Square - fast holewidth measurement with a Doppler spectrometer $\tau_m = 10^{-5}$ s (see Figure 6); point - stimulated photon echo measurement (more exact data [94])

For MAEP-PS at $T = 0.06$ K the holewidth $\delta = 75$ MHz ($\lambda = 617.3$ nm), which exceeds twice the lifetime limit $\delta_0 = 37$ MHz ($\tau_1 = 8.6$ ns). The temperature dependence $\delta(T)-\delta_0$ (Figure 5b) does not display any cross-over points in the range 0.1 - 2 K, $n = 1.12 \pm 0.05$. Extrapolation to $T \to 0$ indicates that in this system the holewidth approaches the limiting lifetime value only at $T \leq 0.01$ K.

The fact that at very low temperatures, $T \simeq 0.1$ K, the holewidth in all the three investigated systems exceeds for several times the lifetime-limited width, approaching the latter only at $T = 0.01 - 0.04$ K, suggests that the impurity interacts with extremely low-frequency excitations of the matrix, amorphous polymer. The thermal properties of polymers in this temperature region are determined by TLS [8,92,93] with the density of states considerably exceeding the phonon density [92]. Consequently, a conclusion can be made that TLS have an actual role in the processes of hole broadening, and that their influence is frozen only at the temperatures T = 0.01 - 0.04 K.

The holewidth in the OEP spectrum is considerably (1.5 - 3 times) larger in PS than it is in PMMA. This result can be understood qualitatively, when using the data about TLS from the measurements of the thermal conductivity of these polymers in the range 0.1 - 1 K [92,93]: the quantity $\gamma^2 N$ (γ - the constant of phonon-TLS interaction, N - averaged TLS density) has a 1.5 times higher value in PS than in PMMA. In the case of MAEP-PS the holewidth at $T > 0.1$ K only slightly exceeds the holewidth in OEP-PS. At $T < 0.1$ K this difference becomes more noticeable, obviously due to the shorter lifetime of the S_1 state in MAEP. A conclusion can be drawn that the holewidth in the temperature region, where $\delta \gg \delta_0$, is determined to a greater extent by the matrix properties than by the properties of the impurity. The latter are more important in the range $T < 0.1$ K, where $\delta \geq \delta_0$, and δ depends strongly on decay processes.

Two regions can be discriminated in the measured dependence $\delta(T) - \delta_0$. At the temperatures $T > 0.1 - 0.2$ K, $\delta(T) - \delta_0 \sim T^{1.1-1.2}$, which is close to $T^{1.3}$-law observed for a number of organic [57] as well as inorganic [20] systems. In the low-temperature region, $T < 0.1$ K, the broadening follows a more steeper dependence $\delta(T)-\delta_0 \sim T^{2.3-2.6}$. Note that such cross-over occurs in a temperature region, where $\delta(T) - \delta_0 \approx \delta_0$, i.e. when the pure phase part of the holewidths[1] becomes comparable to the decay width. Such behaviour can be connected, in principle, with the incorrectness of the subtraction of the decay width from the whole width: the phase and decay parts are additive only in case of the Lorentz shape of the corresponding spectral distributions. However, the decay kinetics of the S_1 state for the systems under investigations is well exponential, hence, the decay contours are Lorentzian. Lorentzian are also hole contours [40] and, consequently, the procedure under discussion is correct.

There is a large number of theoretical studies on the homogeneous ZPL broadening

---

1 It can be supposed that at $T < 0.1$ K, the contribution from the slow spectral diffusion in δ is small, which is indicated by the near-lifetime-limit value of $\delta(T)$ for OEP-PMMA and, consequently, $\delta(T)-\delta_0$ describes the processes of pure dephasing.

in glasses (see Sect. 4) that explain the $T^{1.3}$ relation. However, only a few of them predict one [80] or two [84] cross-overs to a steeper dependence $T^{2.3}$ in the region $T \leq 0.1 - 0.2$ K, considering only the processes of homogeneous broadening during the time $t < T_1$ and neglecting slow processes, $T_1 < t < \tau_m$. At the same time, time-dependent fast measurements in OEP-PS (see Sect. 5.5), depicted in Figure 5a (in the case of holewidth measurement at $\tau_m = 10^{-5}$ s [56,65] and of stimulated photon echo [94]), indicate a considerable influence of slow processes of spectral diffusion on the persistent holewidth, at least at $T \simeq 1.5 - 2$ K. The holewidth is discussed in the model of spectral diffusion in [72]. A broadening by the $T^{4/3}$ relation for dipole-quadrupole (under the condition $N(E) = \text{const}$, E - TLS energy) or dipole-dipole ($N(E) \sim E^{0.3}$) interactions between an impurity and TLS is predicted, whereby the main contribution to it is given by TLS with the relaxation time $T_2 < \tau_r < \tau_m$. However, the important case of very low temperatures, when $T_2' \simeq T_1$ and cross-overs are observed, has not yet been examined theoretically.

Note that on the rise of temperature up to 16 K one more cross-over of the temperature dependence of the holewidth from $\delta_h \sim T^{1.2}$ to $\delta_h \sim T^{1.8}$ is observed at $T = 3$ K [41]. A comparison with LLN data in the region 5 - 11 K, according to which $\delta_f \sim T^3$ and $\delta_h > \delta_f$, shows that alongside with the actual mechanism of homogeneous broadening - a modulational interaction with the quasi-local vibration $\Omega = 20$ cm^{-1} - an additional contribution to the holewidth is made by the process of slow spectral diffusion.

5.5. Microsecond Measurement of Holewidths

Owing to the small quantum efficiency of the hole creation and the "bottleneck" effect in the triplet state, the hole burning + measurement cycle takes $\tau_m = 10^{-2} - 10^3$ s and exceeds considerably the characteristic lifetime of the excited electronic states $T_1 = 10^{-8}$ s. It gives a wide time gap for spectral diffusion. In [65], the experiments with the controlling of τ_m were performed and τ_m down to 10^{-5} s was reached. A sample of OEP-PS was studied in a two-beam scheme: the laser beam was split and a part of it burned the hole, the other part, with the frequency scanned either by Doppler shifts of a rotating mirror [95] or by an acousto-optical modulator, served for detecting the holeshape. The results are presented in Figure 6. The solid curve is the shape of the stable photochemical hole burnt under stationary illumination ($\tau_m = 3$ s). The decrease of τ_m results in a considerable narrowing of the holewidth from $\delta_h = 0.03$ cm^{-1} at $\tau_m = 3$ s to $\delta_h = 0.004$ cm^{-1} at $\tau_m = 10^{-5}$ s. The conclusion is that spectral diffusion is present and causes time-dependent inhomogeneous hole broadening, which may be eliminated at $T = 1.45$ K if τ_m is cut down to the microsecond domain. We should not forget that the spectral diffusion, especially at ultralow temperatures may be accelerated by the hole burning (and even detecting) illumination, as it supplies a considerable amount of the above-the-thermal-background energy selectively to impurity centres. One of the possible mechanisms might be the interaction of the above

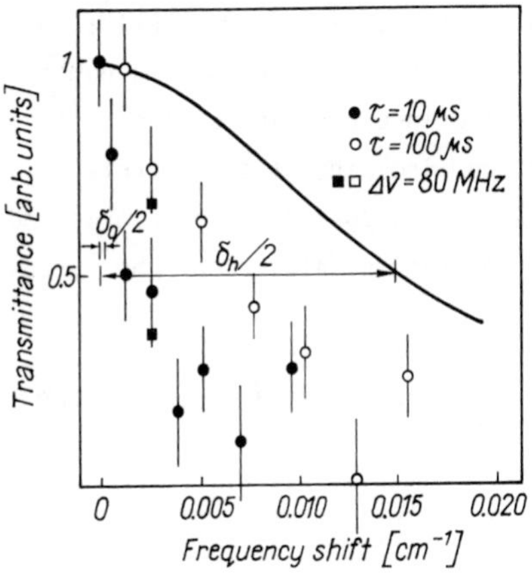

Figure 6: The decrease of the holewidth δ_h in the transmission spectrum of an OEP molecule in PS at 1.45 K with the shortened hole-measurement time τ_m; $\delta_h = 0.03\ cm^{-1}$ at $\tau_m = 3$ s (solid line); $\delta_h = 0.014\ cm^{-1}$ at $\tau_m = 10^{-4}$ s (empty circles); $\delta_h = 0.004\ cm^{-1}$ at $\tau_m = 10^{-5}$ s (filled circles); $\delta_o = 0.003\ cm^{-1}$ - lifetime-limited value. Circles show the data of the Doppler modulation, the squares, the acousto-optic modulation of the laser testing beam

equilibrium phonons (or phonon packets), generated by the relaxing impurity molecule, with the TLS, thus giving rise to the diffusion induced by non-thermal excitations.

6. Conclusion

From the foregoing it follows that a considerable progress, both theoretical as well as experimental, has been achieved in recent years in the investigation of the mechanisms of homogeneous broadening of impurity ZPL in glasses. The fact that in the region of $T < 10$ K the contribution into the broadening is made by low-temperature localized excitations - TLS and impurity QV - may be considered established. In the region of higher temperatures, the direct interaction of the impurity with the intrinsic QV of the glass and with phonons must additionally be taken into account.

A fundamental problem is that of the behaviour of $\delta(T)$ at $T \to 0$. For crystals all the excitations in this interval are frozen, the time of pure phase relaxation $T_2' \to \infty$ and the value of the homogeneous width is determined only by the lifetime of the electronic state $T_1: \delta(0) = (\pi T_1)^{-1}$. Glasses, in contrast to crystals, do not achieve thermodynamical equilibrium even at $T \to 0$. In glasses, transitions to a more stable low-energetic state with the creation of elementary excitations (phonons) is possible. Consequently, if the time of such transition is less than T_1 in the case of LLN or τ_m for SHB, then the values of linewidths (holes), different from the decay limit at $T \to 0$, can be expected. However, up to now there is no experimental confirmation to this situation - the matrix excitations are frozen and they do not broaden ZPL, although it occurs in the region of ultra-low temperatures, $T = 0.1 - 0.01$ K.

Thus, the picture of the homogeneous ZPL broadening in glasses is rather complicated. In view of the diversity of theoretical models on the one hand and the existence of a limited number of reliable experimental data in the considerably narrow temperature region on the other hand, no concrete broadening mechanisms have been elucidated for most of the cases. To make further progress in the understanding of the physics of ZPL broadening in glasses new experimental data are to be extracted by using mutually complementary frequency and time-domain measurements on a number of smaples in a possibly wider temperature region. The incorporation of the information about low-temperature excitations of glasses from independent experiments, such as measurements of heat capacity and thermal conductivity, absorption of ultrasound, neutron scattering, nuclear and electron magnetic resonances, etc. should be purposeful.

The author is indebted to K.Rebane, L.A.Rebane, V.V.Hizhnyakov and J.Kikas for useful discussions and to V.Palm for help with experiments.

References

1 K.K. Rebane: Elementarnaya teoriya kolebatel'noj struktury spektrov primesnykh tsentrov kristallov (Nauka, Moskva 1968) [English transl.: Impurity Spectra of Solids (Plenum Press, New York 1970)]
2 P.M. Selzer, D.L. Huber, D.S. Hamilton, W.M. Yen, M.J. Weber: Phys. Rev. Lett. 36, 813 (1976)
3 A.A. Gorokhovskii, L.A. Rebane: Izv. Akad. Nauk SSSR, Ser. Fiz. 44, 859 (1980) [English transl.: Bull. Acad. Sci. USSR, Phys. Ser. 44, 148 (1980)]
4 A.A. Gorokhovskii, J.V. Kikas, V.V. Palm, L.A. Rebane: Fiz. Tverd. Tela 23, 1040 (1981) [English transl.: Sov. Phys. - Solid State 23, 602 (1981)]
5 W.J. Weber (ed.): Optical Linewidths in Glasses, J. Lumin. 36, N 4/5 (1987)
6 Ch. Kittel: Introduction to Solid State Physics (J. Wiley and Sons Inc., New York 1976)
7 R.C. Zeller, R.O. Pohl: Phys. Rev. B4, 2029 (1971)
8 W.A. Phillips (ed.): Amorphous Solids. Low Temperature Properties, Topics Curr. Phys., Vol. 24 (Springer, Berlin, Heidelberg, New York 1981)
9 P.W. Anderson, B.I. Halperin, C.M. Varma: Phil. Mag. 25, 1 (1972)
10 W.A. Phillips: J. Low Temp. Phys. 7, 351 (1972)
11 L.A. Rebane: in Luminescence of Inorganic Solids, ed. by B.D. Bartolo (Plenum Press, New York 1978) p. 665
12 Yu.V. Denisov, V.A. Kizel: Opt. Spektrosk. 23, 472 (1967) [English transl.: Opt. Spectrosc. USSR 23, 251 (1967)]
13 R.I. Personov, E.I. Al'shits, L.A. Bykovskaya: Opt. Commun. 6, 169 (1972)
14 A.A. Gorokhovskii, R.K. Kaarli, L.A. Rebane: Pis'ma Zh. Eksp. Teor. Fiz. 20, 474 (1974) [English transl.: JETP Lett. 20, 216 (1974)]
15 B.M. Kharlamov, R.I. Personov, L.A. Bykovskaya: Opt. Commun. 12, 191 (1974)
16 C.P. Slichter: Principles of Magnetic Resonance, Springer Ser. Solid State Sci., Vol. 1 (Springer, Berlin, Heidelberg, New York 1980)
17 J. Hegarty, W.M. Yen: Phys. Rev. Lett. 43, 1126 (1979)
18 R.M. Shelby: Optics Lett. 8, 88 (1983)
19 R.M. Macfarlane, R.M. Shelby: Opt. Commun. 45, 46 (1983)
20 J. Hegarty, M.M. Broer, B. Golding, J.R. Simpson, J.B. MacChesney: Phys. Rev. Lett. 51, 2033 (1983)
21 R.T. Brundage, W.M. Yen: Phys. Rev. B33, 4436 (1986)
22 J.M. Hayes, R.P. Stout, G.J. Small: J. Chem. Phys. 74, 4266 (1981)
23 J. Friedrich, H. Wolfrum, D. Haarer: J. Chem. Phys. 77, 2309 (1982)
24 T.P. Carter, B.L. Fearey, J.M. Hayes, G.J. Small: Chem. Phys. Lett. 102, 272 (1983)
25 A. Gutierrez, G. Castro, G. Shulte, D. Haarer: in Organic Molecular Aggregates. Electronic Excitations and Interaction Processes, ed. by P. Reineker, H. Haken, H.C. Wolf, Springer Ser. Solid State Sci., Vol. 49 (Springer, Heidelberg, New York, Tokyo, Berlin 1983) p. 206

26 F.A. Burkhalter, G.W. Suter, U.P. Wild, V.D. Samoilenko, N.V. Rasumova, R.I. Personov: Chem. Phys. Lett. 94, 483 (1983)
27 R. Jankowiak, H. Bässler: Chem. Phys. Lett. 95, 310 (1983)
28 R. Jankowiak, H. Bässler: Chem. Phys. Lett. 101, 274 (1983)
29 H.P.H. Thijssen, A.I.M. Dicker, S. Völker: Chem. Phys. Lett. 92, 7 (1982); H.P.H. Thijssen, R. van den Berg, S. Völker: Chem. Phys. Lett. 97, 295 (1983); H.P.H. Thijssen, S. Völker, M. Schmidt, H. Port: Chem. Phys. Lett. 94, 537 (1983); H.P.H. Thijssen, R. van den Berg, S. Völker: Chem. Phys. Lett. 120, 503 (1985)
30 H.P.H. Thijssen, R.E. van den Berg, S. Völker: Chem. Phys. Lett. 103, 23 (1983)
31 R. van den Berg, S. Völker: Chem. Phys. Lett. 137, 201 (1987)
32 A.A. Gorokhovskii, V.V. Palm: in Sovremennye aspekty tonkostrukturnoj i selektivnoj spektroskopii (MGPI imeni V.I. Lenina, Moskva 1984) p. 69
33 J. Fünfschilling, I. Zschokke-Gränacher: Chem. Phys. Lett. 110, 315 (1984)
34 H.W.H. Lee, A.L. Huston, M. Gehrtz, W.E. Moerner: Chem. Phys. Lett. 114, 491 (1985)
35 L.W. Molenkamp, D.A. Wiersma: J. Chem. Phys. 83, 1 (1985)
36 C.A. Walsh, M. Berg, L.R. Narasimhan, M.D. Fayer: Chem. Phys. Lett. 130, 6 (1986)
37 R. Locher, A. Renn, U.P. Wild: Chem. Phys. Lett. 138, 405 (1987)
38 A.A. Gorokhovskii, V.H. Korrovits, V.V. Palm, M.A. Trummal: Pis'ma Zh. Eksp. Teor. Fiz. 42, 249 (1985)[English transl.: JETP Lett. 42, 307 (1985)]
39 A. Gorokhovskii, V. Korrovits, V. Palm, M. Trummal: Chem. Phys. Lett. 125, 355 (1986)
40 A.A. Gorokhovskii, V.H. Korrovits, V.V. Palm, M.A. Trummal: in Laser Optics of Condensed Matter, Proc. 3rd USSR-USA Symp., Leningrad, 1987, in press; A.A. Gorokhovskii, K.K. Rebane: in Abstracts of the Third International Conference on Unconventional Photoactive Solids, Schloss Elmau, West Germany, Oct. 11-15, 1987, p. 75
41 A.A. Gorokhovskii, V.V. Palm: in Abstracts of the Symposium on Modern Methods of Laser Spectroscopy of Molecules in Low-Temperature Media, Tallinn, May 19-21, 1987, ed. by E. Realo (Estonian SSR Acad. Sci., Tallinn 1987) p. 29; A.A. Gorokhovskii: Izv. Akad. Nauk SSSR Ser. Fiz. 52 (1988), in press
42 V.V. Hizhnyakov: in Teoriya lokal'nykh tsentrov kristalla, Trudy IFA AN ESSR, Vol. 29, ed. by I.J. Tehver (Estonian SSR Acad. Sci., Tartu 1964) p. 83
43 R.H. Silsbee: Phys. Rev. 128, 1776 (1962)
44 D.E. McCumber, M.D. Sturge: J. Appl. Phys. 34, 1682 (1963)
45 M.A. Krivoglaz: Fiz. Tverd. Tela 6, 1707 (1964) [English transl.: Sov. Phys. - Solid State 6, 1278 (1964)]
46 M.A. Krivoglaz: Zh. Eksp. Teor. Fiz. 48, 310 (1965) [English transl.: Sov. Phys. - JETP 21, 204 (1965)]
47 I.S. Osad'ko: Usp. Fiz. Nauk 128, 31 (1979) [English transl.: Sov. Phys. - Usp. 22, 311 (1979)]
48 J.L. Skinner, D. Hsu: Adv. Chem. Phys. 65, 1 (1986)
49 L.A. Rebane: in Ultrafast Relaxation and Secondary Emission, Proc. Int. Symp., Tallinn, Sept. 27 - Oct. 1, 1978, ed. by O.Sild (Estonian SSR Acad. Sci., Tallinn 1979) p. 89
50 K.K. Rebane: Zh. Prikl. Spektrosk. 37, 906 (1982) [English transl.: J. Appl. Spectrosc. 37, 1346 (1982)]
51 L.A. Rebane, A.A. Gorokhovskii, J.V. Kikas: Appl. Phys. B29, 235 (1982)
52 R.I. Personov: in Spectroscopy and Excitation Dynamics of Condensed Molecular Systems, ed. by V.M. Agranovich, R.M. Hochstrasser (North-Holland, Amsterdam 1983) p. 555
53 G.J. Small: in Spectroscopy and Excitation Dynamics of Condensed Molecular Systems, ed. by V.M. Agranovich, R.M. Hochstrasser (North-Holland, Amsterdam 1983) p. 515
54 J. Friedrich, D. Haarer: Angew. Chem. 23, 113 (1984)
55 J. Friedrich, D. Haarer: in Optical Spectroscopy of Glasses, ed. by I. Zschokke (D. Reidel Publ. Comp. 1986) p. 149
56 K.K. Rebane, A.A. Gorokhovskii: J. Lumin. 36, 237 (1987)
57 S. Völker: J. Lumin. 36, 251 (1987)

58 R.A. Avarmaa: Eesti NSV Tead. Akad. Toim. Füüs. Matem. (Proc. Estonian SSR Acad. Sci.) 23, 238 (1974) (in Russian)
59 A.A. Gorokhovskii, J.V. Kikas: Zh. Prikl. Spektrosk. 28, 832 (1978) [English transl.: J. Appl. Spectrosc. 28, N 5 (1978)]
60 J.B. Klauder, P.W. Anderson: Phys. Rev. 125, 912 (1962)
61 J.L. Black, B.I. Halperin: Phys. Rev. B16, 2879 (1977)
62 A.A. Gorokhovskii, V.V. Palm: Pis'ma Zh. Eksp. Teor. Fiz. 37, 201 (1983) [English transl.: JETP Lett. 37, 237 (1983)]
63 W. Breinl, J. Friedrich, D. Haarer: Chem. Phys. Lett. 106, 487 (1984)
64 J. Jäkle: Z. Phys. 257, 212 (1972)
65 K.K. Rebane: Cryst. Lett. Def. and Amorph. Mat. 12, 427 (1985); A.A. Gorokhovskii, V.V. Palm, K.K. Rebane: Eesti NSV Tead. Akad. Toim. Füüs. Matem. (Proc. Estonian SSR Acad. Sci.) (in Russian) to be published
66 A. Szabo: Phys. Rev. Lett. 25, 924 (1970); 27, 323 (1971)
67 M.J. Weber: in Laser Spectroscopy of Solids, ed. by W.M. Yen, P.M.Selser, Topics Appl. Phys., Vol. 49 (Springer, Berlin, Heidelberg, New York 1981) p. 189
68 A. Gorokhovskii, J. Kikas: Opt. Commun. 21, 272 (1977)
69 E.I. Al'shits, R.I. Personov, B.M. Kharlamov: Pis'ma Zh. Eksp. Teor. Fiz. 26, 751 (1977) [English transl.: JETP Lett. 26, N 11 (1977)]
70 A. Szabo: Phys. Rev. B11, 4512 (1975)
71 A.A. Gorokhovskii, J.V. Kikas, V.V. Palm, L.A. Rebane: Izv. Akad. Nauk SSSR, Ser. Fiz. 46, 952 (1982) [English transl.: Bull. Acad. Sci. USSR, Phys. Ser. 46, 116 (1982)]
72 M.A. Krivoglaz: Zh. Eksp. Teor. Fiz. 88, 2171 (1985) [English transl.: Sov. Phys. - JETP 61, 1284 (1985)]
73 D.L. Huber, M.M. Broer, B. Golding: Phys. Rev. Lett. 52, 2281 (1984)
74 S.K. Lyo, R. Orbach: Phys. Rev. B22, 4223 (1980)
75 J.M. Hayes, R.P. Stout, G.J. Small: J. Chem. Phys. 74, 4266 (1981)
76 I.S. Osad'ko: Pis'ma Zh. Eksp. Teor. Fiz. 33, 640 (1981) [English transl.: JETP Lett. 33, 626 (1981)]
77 S.K. Lyo: Phys. Rev. Lett. 48, 688 (1982)
78 B. Jackson, R. Silbey: Chem. Phys. Lett. 99, 381 (1983)
79 S.K. Lyo, R. Orbach: Phys. Rev. B29, 2300 (1984)
80 P. Reineker, H. Morawitz, K. Kassner: Phys. Rev. B29, 4546 (1984); K. Kassner, P. Reineker: Phys. Rev. B35, 828 (1987)
81 T.L. Reinecke: Solid State Commun. 32, 1103 (1979)
82 V. Hizhnyakov, I. Tehver: Phys. Status Solidi b95, 65 (1979)
83 S. Hunklinger, M. Schmidt, Z. Phys. B54, 93 (1984)
84 I.S. Osad'ko: Pis'ma Zh. Eksp. Teor. Fiz. 39, 354 (1984) [English transl.: JETP Lett. 39, 426 (1984)]; Chem. Phys. Lett. 115, 411 (1985); Zh. Eksp. Teor. Fiz. 90, 1453 (1986) [English transl.: Sov. Phys. - JETP 63, 851 (1986)]; I.S. Osad'ko, A.A. Shtygashev: J. Lumin. 36, 315 (1987)
85 A. Alexander, C. Laermans, R. Orbach, H.M. Rosenberg: Phys. Rev. B28, 4615 (1983)
86 R.M. Macfarlane, R.M. Shelby: J. Lumin. 36, 179 (1987)
87 M.M. Broer, B. Golding, W.H. Haemmerle, J.R. Simpson, D.L. Huber: Phys. Rev. B33, 4160 (1986)
88 A.P. Marchetti, M. Scozzafava, R.H. Young: Chem. Phys. Lett. 51, 424 (1977)
89 J.M. Hayes, G.J. Small: Chem. Phys. Lett. 54, 435 (1978)
90 K.N. Solov'ev, I.E. Zalesski, V.N. Kotlo, S.F. Shkirman: Pis'ma Zh. Eksp. Teor. Fiz. 17, 463 (1973)[English transl.: JETP Lett. 17, 332 (1973)]
91 A.A. Gorokhovskii, V.V. Palm: in Abstracts of the Symposium on Modern Methods of Laser Spectroscopy of Molecules in Low-Temperature Media, Tallinn, May 19-21, 1987, ed. by E. Realo (Estonian SSR Acad. Sci., Tallinn 1987) p. 27
92 R.B. Stephens: Phys. Rev. B8, 2896 (1973)
93 J.E. Graebner, B. Golding, L.C. Allen: Phys. Rev. B44, 5696 (1986)
94 R.K. Kaarli, M.L. Rätsep: in Abstracts of the Symposium on Modern Methods of Laser Spectroscopy of Molecules in Low-Temperature Media, Tallinn, May 19-21, 1987, ed. by E. Realo (Estonian SSR Acad. Sci., Tallinn 1987) p. 39
95 K.K. Rebane, V.V. Palm: Opt. Spektrosk. 57, 381 (1984) [English transl.: Opt. Spectrosc. USSR 57, 229 (1984)]

Zero-Phonon Lines and Time-and-Space-Domain Holography of Ultrafast Events

P. Saari

Institute of Physics, Estonian SSR Acad. Sci.
SU-202400 Tartu, Estonian SSR, USSR

Abstract

The physical nature and features of space-and-time-domain holograms are discussed. A review of theoretical and experimental results on the space-and-time-domain holography is given (see also [15]). Problems of taking into account polarization coordinates in the procedure are discussed and possibilities to play back time-dependent fields of arbitrary polarization are shown.

1. Introduction

In a monograph by REBANE [1], the noteworthiness of the vanishingly small width of zero-phonon lines (ZPL) as a physical phenomenon has repeatedly been emphasized and their perspectives in spectroscopic investigations discussed. Indeed, a number of fine spectroscopic effects observed subsequently proceed from the same remarkable ability of an impurity electronic oscillator to have in a condensed matter with inherent strong interactions an exceptionally high Q-factor limited only by the losses through interaction with electromagnetic vacuum [2,3]. The radiative width of ZPL (10^8 s^{-1} and less) is by 3-4 orders of magnitude less than the typical vibrational relaxation rates in a crystal, by 4-6 orders less than the phonon frequency and approximately by the same amount less than the frequencies corresponding to the binding energies of particles in a crystal. Therefore, ZPL serve as extremely sensitive and precise mediators of the information about interactions in an impurity centre - by means of light they provide a comprehensive spectroscopic information about crystals. Such spectroscopic application has to be extremely fruitful in the investigations of the fundamental problems of the interaction of light with impurity systems (e.g. classification of secondary emission and discovery of hot luminescence, see review [4]) as well as in the studying of particular impurity crystals and in the detection of various fine mechanisms of microdynamics in them (see review [5]). The possibilities of spectroscopic applications of ZPL have grown considerably with the discovery of the effect of photochemical hole burning [6] (PHB), that enables one to avoid the inhomogeneous broadening unfavourable from the spectroscopic point of view.

However, PHB is far from being only a method of spectroscopic investigations. Its essence lies in the possibility of forming optical characteristics of the matter by means of light as an instrument: highly-selective changing of the absorption coefficient, consequently, also the refractive index, i.e. the tensor of dielectric permittivity in general, in spacial as well as frequency domain. In addition, thanks to PHB, the extra high Q-factor of ZPL can be utilized also in storing the information carried by light. At this point the inhomogeneous broadening turns into a useful property of the medium - the larger it is, the "longer" is the sample in the fourth, the frequency dimension, the more spacious are the "places" for disposing information. The ensuing prospects of building optical memories, in which the scanned monochromatic laser radiation first burns holes at different frequencies, then reads them, i.e. bits up to 10^5 units per surface element of the dimensions λ by λ, have been discussed in literature ever since the first investigations of PHB appeared (see review [3]).

However, the exceptional properties of the media with ZPL, prone to PHB, allow one to go further and set the task of information storage-reading in the following maximal way: to record the characteristics of the light field in the given four-dimensional region completely in the medium and then to make the medium to play this field back at any desired moment in the form it was stored, in all its spatial and temporal dependences.

Such task has been solved starting from the initial ideas and their experimental verification and ending with the demonstrations of practical applications in holographic processing of events [7-15]. To give at once an idea of what should be understood by time-and-space-domain holography - this is what the main concern of the present work will be - and what scale of the physical parameters will be discussed, one of the most recent experimental results by R.Kaarli and A.Rebane is presented in Figure 1.

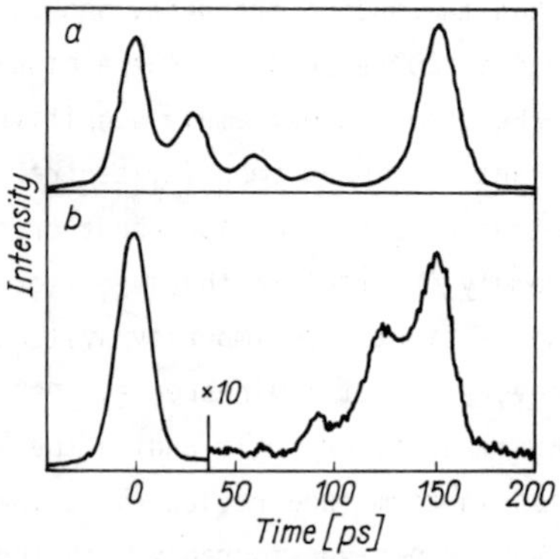

Figure 1: Play-back of the signal in the temporally reversed form. a - amplitude-modulated picosecond signal to be recorded and the reference pulse, b - read-out pulse directed onto the hologram within some tens of minutes after the recording procedure and time-reversed replica of the signal. Pulse profiles have been recorded by a streak-camera with temporal resolution 20 ps

This as well as some other results from [7-15], in particular those about ZPL peculiarities,thanks to which the medium is able not only to memorize but also to reverse the temporal behaviour of the signal, will be analyzed in the main part of the text.

The aim of the present paper, along with the presentation of the physical essence and properties of time-and-space-domain holograms, is to discuss their interrelations with the physical nature and characteristics of ZPL and PHB. It will be shown that the high Q-factor of electronic oscillators enables the realization of not only time-and-space-domain holography as such but in conjunction with the oscillators' immobility in space. It is the latter circumstance that gives one of the aspects differentiating ZPL from narrow lines in gases.

Impurity oscillators are distributed in the matrix not only along the spatial and frequency coordinates but also over the orientations of the dipole momentum of the electronic transition. This indicates the possibility of recording not only the spatial and temporal, but also the polarizational characteristics of the light field, including the time dependence of its polarization state. We are going to demonstrate the theoretical possibility of incorporating polarizational coordinates into time--and-space-domain holography.

Thus, thanks to the media with ZPL and PHB, a prospect will be opened to such holography which ultimately corresponds to the word-for-word meaning of the term - "full recording".

This paper will not concern the account of the theory and methods of time-and--space-domain holography, the experimental results with signals reflected from real objects, the survey of literature on the related fields mentioned (dynamical holography, nonlinear coherent transients) - all this can be found in reviews [14,15]. Here, instead, the following problems will be analyzed in detail:

1. How is the information about the temporal dependence of the light wave preserved in a medium?
2. Why is the phase information not erased by phonons?
3. Why is it a solid medium with ZPL that allows the realization of time-and--space-domain holography?
4. What is the physical nature of the emission responsible for the play-back of the signal?
5. In what sense is the holographying process linear and in what sense nonlinear with respect to the interaction of the field with the matter?
6. What is the practically-applicable limit diffraction efficiency of these holograms?
7. What role is played by the peculiarity of the refractive index, accompanying the absorption ZPL?
8. What restrictions are set by the spectral characteristics of the medium to the temporal characteristics of the holographed events?
9. What perspectives are opened by time-and-space-domain holography in optical data processing?

2. Spectral Hologram and Modulation of the Dielectric Permittivity of the Medium

In this chapter, the mechanism of memorizing the temporal dependence of a light signal is regarded in terms of the spectral language. In spectral representation, the main idea of time-domain holography and its analogy with the principle of recording phase information in ordinary holography are more transparent. Moreover, ZPL is just a term of the spectral language.

Let there be a plate of a photosensitive material which, after some exposure to light, not simply changes its transparency, but does it spectrally selectively. Let thereby either growth or deterioration of transparency at any wavelength be proportional to the spectral intensity of the incident light. If prior to the exposure the transmittance of the plate was independent of frequency - the plate was "gray", then after it the transmittance becomes spectrally modulated, whereby the frequency dependence corresponds strictly to the spectrum of the acting light if the modulation is shallow enough. A number of frozen dye solutions subject to PHB serve in good approximation (in which, it will be discussed below) as highly-selective photochromic media. Evidently, the PHB effect itself is a special case of recording the spectrum of light in case the latter is monochromatic.

Thus, after the exposure the plate starts operating as a frequency filter whose transmittance coefficient acquires a frequency-dependent multiplier, $K(\omega)$. If a "read-out" beam of light of any spectral composition is now directed onto the plate, then, provided the incident field is weak enough and nonlinear optical effects are respectively absent, this composition, on passing through the plate, is multiplied by $K(\omega)$. Naturally, the modulation $K(\omega)$ is not long-lived: the "read-out" light starts to change it according to its own spectral composition. However, in practice it is possible to suppress this effect either by restricting the total energy of the "read-out" light or by using the PHB process occurring only in the presence of a gating beam in another frequency range (see reviews on PHB in this issue).

If now a light pulse is directed to the plate, which is so short that its spectrum uniformly embraces all the dependence $K(\omega)$, then at the output this initially "white" pulse will evidently have the spectral composition $K(\omega)$. What does the time-dependence of the field look like at the output? The answer to this question comes from the well-known relation between the responses of a linear filter: temporal response is determined by the inverse Fourier transform of the causality-corrected spectral response function $K_C(\omega)$, that in the present case gives $Y(t-t_0)\cdot A(t-t_0)$, where t_0 is the moment of the arrival of the read-out pulse at the given point behind the plate, Y is the Heaviside unit step function providing the cut-off of "the future" ($Y=1,\ t \geq t_0,\ Y=0,\ t<t_0$) and A is the autocorrelation function of the "recorded field" on the exposure. We can see that a remarkable effect is predicted by elementary considerations: the short pulse (say, a picosecond pulse, then its length toward propagation is of the order of $\sim$1 mm), which passes through the plate, acquires a "tail" whose length may considerably exceed that of the incident pulse and

whose duration is by no means limited by the transit time of the pulse through the whole plate thickness. The "tail" is not caused by a sort of reflection from plate surfaces either, as it is, for insatnce, in the Fabry-Perot interferometer. What is the physical nature of this "tail" as a response of the matter to light, considering that it arises at however weak incident field and propagates in one direction only? Evidently, the physical reason of the appearance of the "tail" can be nothing but the pulse-induced coherent polarization of the electronic oscillators of the medium, responsible for selective absorption. Thereby, as their frequency distribution is not "white" but is determined by the spectrum of the recorded light, the resulting macroscopic polarization decays according to the function A(t). In other words, it is the same linear polarization, which in case of a monochromatic wave extinguishes the wave in the medium propagating with the velocity c, and replaces it by a wave propagating slowlier in accordance with the dispersion properties of the resonant medium. In terms of the resonant secondary emission of impurity molecules, the component responsible for the delayed response can be classified as a "forward Rayleigh scattering". The inconsistent content of this term is due to the circumstance that the description of the secondary emission of impurity systems, developed so far (see V.V.Hizhnyakov's references in this issue), does not allow for the counteraction of impurity centres on the field. It should be mentioned that in connection with the rapid development of the spectrochronographic investigation of the components of the resonant secondary emission of condensed media the elaboration of a self-consistent theory of secondary emission is quite actual.

Finally, if one asks what restricts the limit duration of the response under consideration, then one must again turn to the same high Q-factor of impurity oscillators and the answer would be evident - the possible duration of the response is determined by the reciprocal of the homogeneous width of the ZPL.

However, the autocorrelation function cannot determine uniquely the temporal behaviour of the recorded field. For instance, it may be quite identical for a long-lasting noise signal on some average optical carrier frequency and for a short pulse signal of a certain form on the same carrier.

It is evident that for the response of the medium to play back the temporal behaviour of the recorded field it is necessary to organize the storage so that $K(\omega)$ should contain the Fourier transform of the signal and not only its spectrum. To do this, one must resort to a common holographic procedure allowing one to do without the indispensable phase loss on recording the amplitude of an optical field. By the way, this procedure serves as the basis for a number of spectroscopic methods (of nonlinear polarizabilities, excitation profiles of the Raman scattering, etc.). The essence of the procedure lies in the following: to the complex number S to be recorded a certain "reference" number, for example, one, is added. Then, for the square of the modulus of the total value we obtain: $|S+1|^2 = 1 + |S|^2 + S + S^*$. In our case, if we use the complex field representation, in the capacity of the value S serves the Fourier transform of the signal $S(\omega)$. The idea of the procedure will be the same when describing the field and its Fourier transform via real trigonometric functions, but the mathematical relations will be somewhat more complicated. We can see that by means of the third (with some reservations also by the fourth) term the recording takes place without any phase loss. It is understandable that, to avoid unnecessary complications, the reference field should be selected so that its Fourier transform be, in a certain sense, alike for all the frequencies contained in the signal. It is purposeful to choose the reference Fourier transform with the frequency dependence $\exp(-i\omega t_d)$, the modulus of which is constant, similar to $1(\omega)$, whereas t_d is some delay, which has the same sense for time-domain holography as the derivation of the reference wave has in ordinary holography. If the reverse Fourier transform is taken with the start of time account at the point of the forefront of the signal field, it can be seen that our reference field must be a δ-shaped pulse delayed from the signal by the time interval t_d (see Figure 2a). Being δ-shaped means that the spectrum of the reference pulse should cover that of the signal.

After exposure the plate again turns into a frequency filter, its transmittance coefficient being now determined by the spectrum of the pair of the signal and reference pulses:

$$K(\omega) \sim 1 + |S(\omega)|^2 + S(\omega)\exp(i\omega t_d) + S^*(\omega)\exp(-i\omega t_d) \quad . \tag{1}$$

In other words, in the absorption spectrum, by means of pulsed spectral hole burning, a spectral hologram of the signal has been memorized. If the contrast of the hologram burnt by the pair of pulses is not sufficient, it can be enhanced by a multiple repetition (m times) of the pairs at arbitrary time intervals t_m. Thereby, however, the following requirements should be fulfilled: (i) the delay of the reference pulse should be repeated with the accuracy up to tenths of the period of the optical carrier, (ii) the time dependence of the signal phase should be repeated with the same accuracy, (iii) on the contrary, t_m should be distributed randomly, so that on averaging the cross terms of the spectrum with the multipliers $\exp(\pm i\omega t_m)$ should turn to zero.

In reality, the last requirement is removed when the intervals between the pairs exceed the centres' phase relaxation time, i.e. the reciprocal of the homogeneous width of the ZPL. Naturally, in case of too large m, spectral holes will be saturated and the terms nonlinear with respect to $S(\omega)$ appear in (1). The corresponding additional responses, however, are weak and easily avoidable in the experiment (see below).

Now the last problem is to be considered: what does the temporal response of the hologram to the read-out pulse look like? As before, the response is given by the reverse Fourier transform, $K(\omega)$, multiplied by the causal function $Y(t-t_0)$. Thereby, the first term in (1) gives the passed-through undistorted read-out pulse[1], the sec-

[1] More exactly, in case of a finite duration of the reference and read-out pulses - double autoconvolution of the pulse.

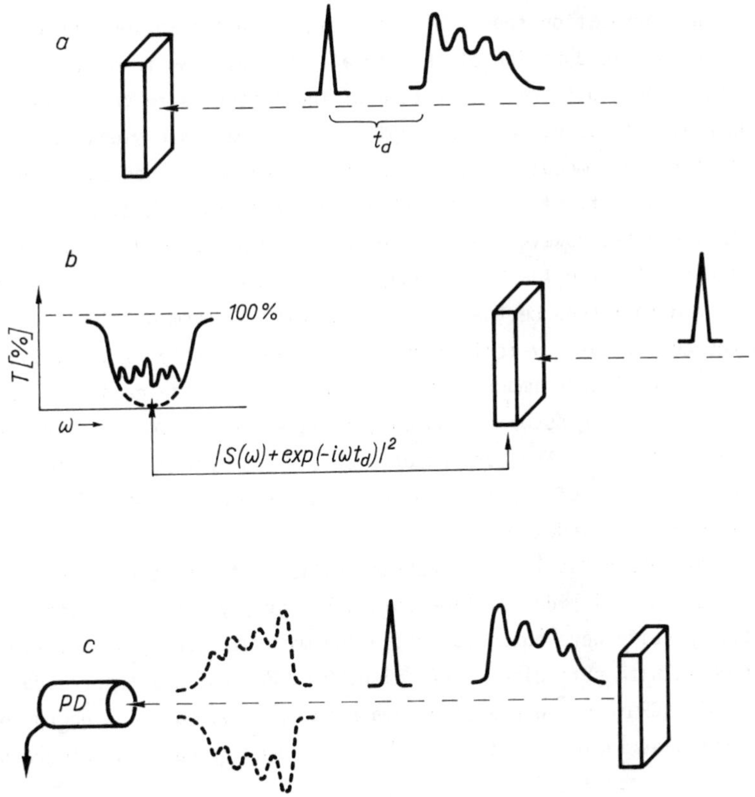

Figure 2: Process of time-domain holography. a - recording: by pairs of reference and signal pulses, delayed from each other by the interval t_d, a highly-selective photochromic medium is irradiated in which, as a result of the effect of pulsed photochemical hole burning, a spectral hologram is memorized; b - the plate serves as a linear filter for the read-out pulse; the frequency dependence of filter transmittance is shown, which in the region of the inhomogeneous absorption band exhibits the spectral hologram; c - the hologram response to the read-out pulse, recorded by an ultrafast light detector. The broken line indicates responses arising through the terms $S^*(\omega)$ in absorption and dispersion; as these responses violate the causality principle, they annihilate each other

ond one, the autocorrelation function of the signal, whose maximum duration is of the order of the double duration of the signal. The last two terms in (1) are of most interest. According to the first term of the latter, with the delay t_d with respect to the read-out pulse, an exact replica of the recorded signal is played back, including its frequency modulation, in case the signal contained one[2]. The last term indicates the possibility of playing back a replica of the signal with reversed temporal course of events, however, as it should outstrip the read-out pulse, the causality function Y cuts it off in this case. In Chap. 4, we shall see that there is really no need to introduce the causality condition "from outside". It is automatically introduced by the physical nature of the plate's response, namely, the polarizational response of each impurity oscillator satisfies the causality condition in a natural way. In the macroscopic language, it is expressed in a really complex transmission of the plate, i.e. the burning of a spectral hologram in the absorption spectrum implies the burning of the corresponding hologram also in the frequency dependence of the refraction index.

At the end of 1981 R.Kaarli and A.Rebane started with the experiments to check the above-discussed basic idea of time-domain holography. It took a year before the first positive results were obtained, while the main difficulty at the time lay in the lack of media with sufficinetly efficient PHB. Meanwhile a theoretical paper was published by MOSSBERG [16], where time-domain holography, based on photon echo in inhomogeneous resonant media, was proposed. Later he performed such holography experimentally on gaseous media [17,18].

As can be seen in Figures 3,4, reproduced from the first investigations [7-9], the experiment proves the practical realizability of time-domain holography.

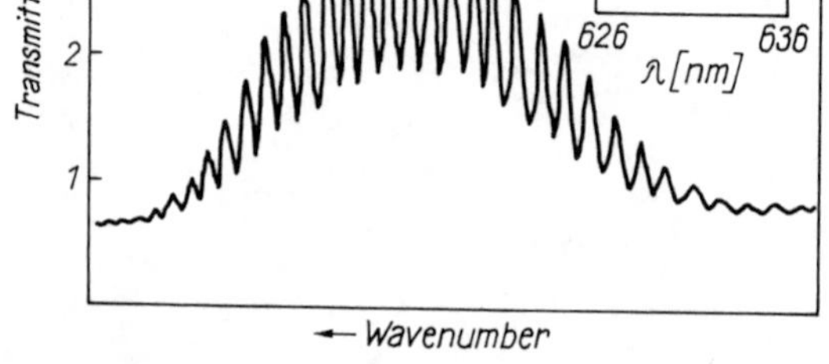

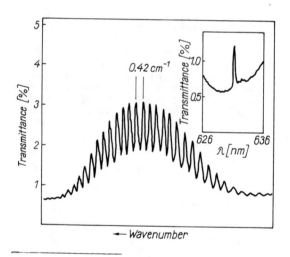

Figure 3: Spectral hologram in the transmittance spectrum of H_2-tetra-tret-butyl-porphyrazine in polystyrene after an exposure by picosecond pulse trains. In the insert - a general view of the hole measured with poor resolution (∿0.1 nm)

2 The restored signals are actually also convolutions with reference and read-out pulses; however, if the duration of the latter is less than that of the shortest peculiarities in the signal, this specification is unimportant.

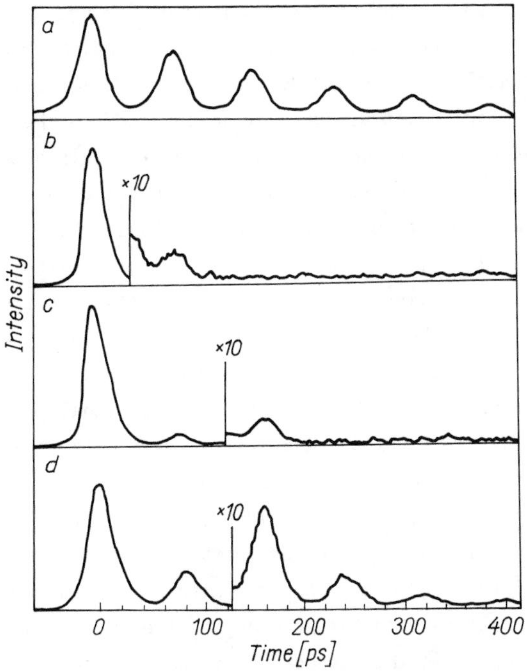

Figure 4: Dependence of the response of octaethyl-porphin in styrol on the exposure at photochemical burning by picosecond pulse trains with the interval 80 ps (a); responses after exposure of 0.5 (b), 1.5 (c), and 2.5 mJ/cm^2 (d). It can be seen that the response, appearing after the read-out pulse (at t = 0, curve d), reconstructs the memorized signal (curve a). In comparison with the schematic in Figure 2, here the separated reference pulse is absent, one can take for it the first pulse on the curve a

To conclude this chapter, note that time-domain holography is fully analogous to the ordinary Fourier holography of a one-dimensional image. In the latter case, light-sensitive centres along the transverse coordinate from end to end of the hologram record the spatial frequencies, in the former case, along the frequency coordinate, from end to end of the inhomogeneous absorption band they record the temporal frequencies of the signal. An essential difference of a time-domain hologram, however, lies in the asymmetry of reproduction - the image of the signal is not played back in the region $t < 0$ because of the causality condition.

3. Phase Memory of a Medium with ZPL and Pulse Response of a Spectral Hologram

The process of holographying events has two substages. First, the medium containing impurity molecules with inhomogeneous distribution of resonance frequencies, subject to photochemical transformation, is repeatedly irradiated by a picosecond exciting signal until a frequency structure corresponding to the signal appears on the absorption band, i.e. the stage of recording the hologram by means of the pulsed photochemical hole burning phenomenon, PSHB [7-9,13]. After that, in tens of minutes or hours, the irradiated medium is probed by a single picosecond pulse, after transmission of which the medium radiates a coherent delayed response in the form of a replica of the signal - this is the stage of the read-out of the hologram by means of the phenomenon of photochemically accumulated stimulated photon echo (PASPE) [7,9,13].

Let us regard now in time representation how and by what is the memorizing of phase relations between the signal and reference pulses achieved in the medium. Here, evidently, two entirely different time scales are involved. The first of them is connected with the duration of the signal (more exactly, here the interval up to the reference pulse is also added). The second one is connected with the intervals between the acts of storage and read-out of the hologram. On the first one, the recording (as well as restoring) of the phase information, on the second, its storage, take place. Let, for clearness, the signal (not only with temporal but also with spatial modulation) be ended with a δ-shaped pulse with the wavefront of an arbitrary spatial configuration. On passing through the recording medium of the (plane) reference pulse, the electronic oscillators distributed over the medium volume are alternately "jerk-wise" excited. For the oscillators to fix correctly the signal phase in each spatial point, i.e. in particular, to count the time $\tau(\underline{r})$ between the moments of passing through any point $\underline{r}$ of the reference wavefront and the wavefront at the end of the signal, the behaviour of the oscillators should satisfy the following conditions. First, in this time, their position in space must remain unchanged with the accuracy of the order of a tenth of the wavelength. Second, within the same time they must retain the phase of their vibrations with the accuracy of the order of a tenth of the period. The atoms and molecules in the gaseous phase, having the homogeneous widths of optical resonances $\leq 10^9\ s^{-1}$ and the rates of thermal motion of $\sim 10^2$ m/s, satisfy these requirements in the nanosecond time scale. However, for these the other time scale - the scale of the longevity of preserving the hologram - also consists of nanoseconds, as at larger times there occurs the smearing of the spatial distribution of excited oscillators and after that, due to energetic relaxation, the information about the time dependence of the signal disappears as well. The latter dependence has been fixed earlier in the population of the resonance level in dependence on the Doppler distribution of the resonance frequency of molecules: in this case, the oscillations with the frequencies $\omega = 2n\pi/\tau(\underline{r})$ took place in phase with the field of the signal pulse and, consequently, the probability for these molecules to pass over to a higher level was maximum (conversely, for molecules with $\omega = (2n+1)\pi/\tau(\underline{r})$ it was minimum). As a result, an interference picture of the form $1+\cos\omega\tau$ is formed on the ensemble of oscillators along the frequency coordinate; as it should be, the spectral intensity of two pulses with the interval τ contains the same multiplier. It is interesting to note that in the frequency space also the fields interfere, which do not yield ordinary interference [14].

Much more advantageous is the situation with impurity molecules or centres in low-temperature solid media. Here the spatial immobility is ensured with the precision up to the thousandths of the wavelength, the PHB, however, by means of destroying (or by transfer into an unactual resonant frequency region) the molecules, which with some probability remain excited, memorizes the hologram in the spatial-frequency distribution density of molecules. Thus, for such information-recording media the second scale runs up to tens of hours and possibly months. On the other hand, at the first glance, the situation with the phase-preserving time is rather poor. Owing to electron-phonon interaction, the vibrations of the surrounding disorderly shake the resonance frequency of the (electronic) oscillator. For a typical width of the phonon wing the corresponding correlation time is not more than 10^{-13} s. However, besides the wing, there exists a much narrower ZPL[3]. This indicates the presence of some constant constituent in the random process of modulation of the electronic transition frequency by phonons. The thing is that although at linear electron-phonon interaction the vibrations of the medium rather strongly shake the frequency of the electronic oscillator, on the average the phase of the latter is preserved. The reason of preserving the phase is the following: the spectrum of a chaotic modulation, i.e. the spectrum of phonons in a three-dimensional medium, turns to zero when the frequency of phonons approaches zero, consequently, in the modulation process there are no components able to carry the frequency of oscillators away for an unlimited time and thus to bring about the phase diffusion [19].

As a result we can see that on recording, like in case of gaseous media, the phase memory is determined by nanosecond or larger times of either the radiational or radiationless decay of the excited states of the oscillator and by other processes of the homogeneous broadening of ZPL. The presence of phonon wings leads to the formation of some background under the hologram, which lowers the contrast of the latter.

On restoring, the read-out pulse, in its turn, excites electronic oscillators. For the macroscopic polarization to be able to play back the signal, the oscillators of the same frequency in each point of the medium should vibrate in phase at least until the end of the formation of the signal. As the response of such oscillators is given by the expression

$$P(t) \sim Y(t)\exp(-i\omega t)\exp(-2t/T_2) \quad , \qquad (2)$$

we can see that the maximum duration of the play-back signal is again determined by the reciprocal of the ZPL width, i.e. by the phase relaxation time T_2.

As a result, we obtain the following relations between the characteristics of the medium and the holographed signal: (i) the duration of the signal is limited by the phase relaxation time of the zero-phonon transition (typically ∿10 nanoseconds); (ii) time resolution is limited by the time of reversible phase relaxation, i.e. the reciprocal of the inhomogeneous width of the resonance band (typically 10 - 100 femtoseconds); (iii) the longevity of the hologram is limited by the lifetime of PHB products (in the dark and at liquid helium temperatures up to tens of hours).

3 There is a concept that the ZPL is the result of a quantum-mechanical approach. Actually it is vice versa: at $T=0$ the spectrum of a classical impurity crystal would contain only a ZPL, the phonon wing, however, is the result of zero-point vibrations of the lattice, i.e. at low temperatures the wing is a purely quantum effect.

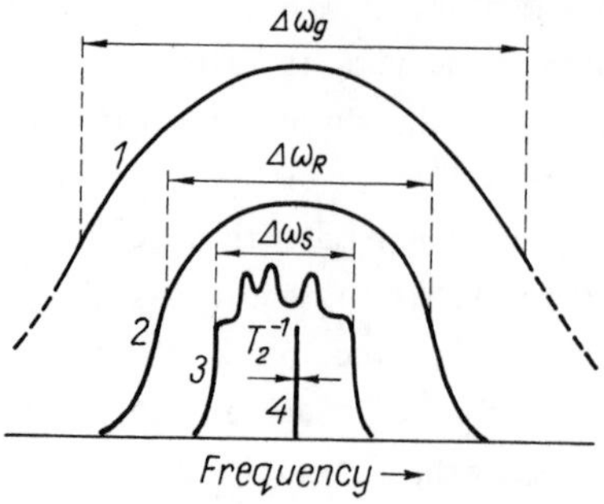

Figure 5: Relations between the spectral widths of the inhomogeneous absorption band (1), the reference pulse (2), the signal (3), and ZPL (4)

The corresponding frequency relations are depicted in Figure 5.

At the beginning of the chapter, the effect of the formation of the delayed coherent macroscopic response was called PASPE. As one could see, the consideration of the physical nature of this effect in temporal representation really reminds of the physics of a three-pulse (stimulated) photon echo (PE). This analogy has been considered in more detail in [13], where the appearance of PASPE is described in terms of the pseudospin vector model. From the comparison of these two effects the following can be inferred:

1. They are similar in the sense that by the moment of applying the read-out pulse an irreversible phase relaxation has taken place and the information about the exciting pulses is contained in the spectral modulation of the absorption coefficient of the medium (in case of PE, due to the modulation of population differences, for PASPE, due to the modulation of the number of molecules).
2. For both of them the delay time τ must be shorter than that of irreversible phase relaxation.
3. In case of PASPE the role of the energetic relaxation time T_1 is played by the practically unlimited lifetime of photoproducts, therefore, PASPE can be obtained repeatedly at arbitrary time moments.
4. The coefficient of the transformation of the read-out pulse energy into the PASPE-response can be high at arbitrarily small areas of all three pulses if the action of the first two has been accumulated by the repetition of burning cycles for a sufficient number of times.
5. The formation of PASPE does not require nonlinearity typical of a two-level quantum system. All its formation stages are describable within the frames of linear oscillators without anharmonicity (this is the difference between the PASPE and the oscillator echo), and the total *linear* response of their ensemble is what forms the echo signals.
6. The role of nonlinearity, necessary for the formation of echo phenomena, is played by the ability of excited oscillators to decay. The nonlinearities with respect to spectral intensity of the pair of exciting pulses, arising due to

the saturation of the PHB process, lead to the formation of repeated responses delayed by $n\tau$ $(n=1,2,\ldots)$, thereby the signal is autoconvoluted by n times [20].

7. There is a phenomenon intermediate between PASPE and PE - accumulated (at the triplet state) stimulated photon echo [21].

4. Combination of Time- and Space-Domain Holographies and the Possibilities of Holographying Vector Fields

In ordinary holography, the recording of the amplitude $S(\underline{r})$ of the field scattered from the object is obtained by adding an amplitude $R\cdot\exp(i\underline{K}_R\cdot\underline{r})$ to $S(\underline{r})$ in a photosensitive medium; here $\underline{K}_R$ is the wave vector of the plane reference wave with the intensity R^2. The reference wave is extracted from the monochromatic laser beam illuminating the object and directed to the hologram at some angle Θ toward the mean direction of the object wave. The tilt and spacing of interference fringes are determined by Θ; in dependence on the ratio of the spacing to the thickness of the hologram the latter are classified into thin and thick holograms (Denisyuk's hologram). These two types of holograms possess some distinctive features, e.g. a thin hologram always restores two images - virtual and real - with the angle 2Θ between them; the restauration by a thick hologram is governed by the Bragg condition and gives only one - either virtual or real - image depending on the direction of the read-out. Under red light and with $2\Theta \simeq 4^0$ the holograms with the thickness less than 0.1 mm behave like the thin ones.

As the spatial resolution of the media with PHB is not inferior to that of holographic photomaterials, on the contrary, it approaches the values determined by the dimensions of impurity molecules, their employment in holographying the spatial dependences of the field is quite potential. The presence of the frequency coordinate, however, typical of these media, is used to record the temporal behaviour of the field. As a result, in such media a time-and-space-domain holography is possible, i.e. the recording of scene images with a subsequent play-back of them in their full temporal development (motion of the parts of the object scene, the arising and decay and phase modulation of the light sources in the picosecond time scale). Dynamical holography, too, is concerned with nonstationary fields. To avoid a terminological confusion note that holography is called dynamical in the media where the formation of the hologram and the transformation of the read-out waves on it take place in "the real time" [22]. In the limiting case, a dynamical hologram follows the spatial distribution of an instantaneous light intensity in the interference picture. The time-and-space-domain hologram, however, records (for a time considerably exceeding the duration of the event recorded) the whole dependence of the distribution mentioned.

The theory of time-and-space-domain holography based on the PSHB and PASPE has been elaborated in [10] (an improved version in [20]), the corresponding experiments are described in [11,12]. Here we shall refer to the theory and give a short characterization of the corresponding experimental results (see also [14,15]).

The signal is described by a complex amplitude $S(\underline{r},t)$, whose dependence on its arguments is slow as compared to the exponent of the optical carrier, $\exp[i\omega_o(t-z/c)]$; it is assumed that the signal wave of the duration t_S and average frequency ω_o propagates along the z-axis and at the time moment $t=0$ is incident on the recording plate in the plane (x,y). At the moment t_R a δ-shaped reference pulse of plane waves with the amplitude R_0 falls onto the same plate with the propagation direction at an angle θ with respect to the z-axis and the characterizing vector $\underline{n}_R = c^{-1}(-\sin\theta, 0, \cos\theta)$. It should be emphasized that, in contrast to the ordinary holography, here the reference wave is nonstationary; by matching the delay time t_R the reference and signal pulses can be (and generally it is necessary) separated in time (see Figure 6a) so that their usual interference does not appear in the hologram. The requirement of the coherence of the signal and reference waves on performing data-storage with only one pair of pulses falls off as well.

In the point $\underline{r}$ of the recording plate, the impurity molecule with the resonant frequency ω "feels" the square of the modulus of the Fourier component of the field with the frequency ω (cf. formula (1)):

$$\bar{E}\bar{E}^* = R_0^2 + |\bar{S}(\underline{r},\omega-\omega_o)|^2 + R_0\bar{S}(\underline{r},\omega-\omega_o)\exp[-i\omega(\underline{n}_R\cdot\underline{r} - t_R)] + R_0\bar{S}^*(\underline{r},\omega-\omega_o)\exp[+i\omega(\underline{n}_R\cdot\underline{r} - t_R)] \,, \tag{3}$$

where $\bar{S}$ is the Fourier transform of the signal. Molecules are excited via absorption with the cross-section δ and then annihilated via PHB with the probability η. Hence, the initially homogeneous density g_0 of the distribution of molecules over the spatial and frequency coordinates is changed in the course of hologram storage:

$$g(\omega,\underline{r}) = g_0(1 - k\bar{E}\bar{E}^*) \,, \tag{4}$$

where $k = \delta\eta m$, and m is the number of pulse pairs during the whole exposure. Expression (4), describing the formation of a spatial-spectral lattice-hologram, holds only for a small burning depth, $k\bar{E}\bar{E}^* \ll 1$. In conformity with (2), a molecule with the resonance frequency ω' has the frequency dependence of susceptibility $\sim[\omega(\omega' - \omega + i/T_2)]^{-1}$. The total susceptibility χ of the medium is given by integration over resonance frequencies ω' with the distribution $g(\omega',\underline{r})$. At large T_2 (see Figure 5) this integration is reduced to the convolution of the Heaviside step function with the Fourier transform, manifested via $(1+i\hat{H})$, where $\hat{H}$ is the Hilbert transform operator acting upon the frequency function standing behind it. As the dielectric permittivity $\varepsilon = 1 + \chi$, then for the change of ε at the expense of the hologram storage we obtain from (4):

$$\Delta\varepsilon(\omega,\underline{r}) = i(\delta c/2\omega)k(1+i\hat{H})\{\bar{E}\bar{E}^*\} \,. \tag{5}$$

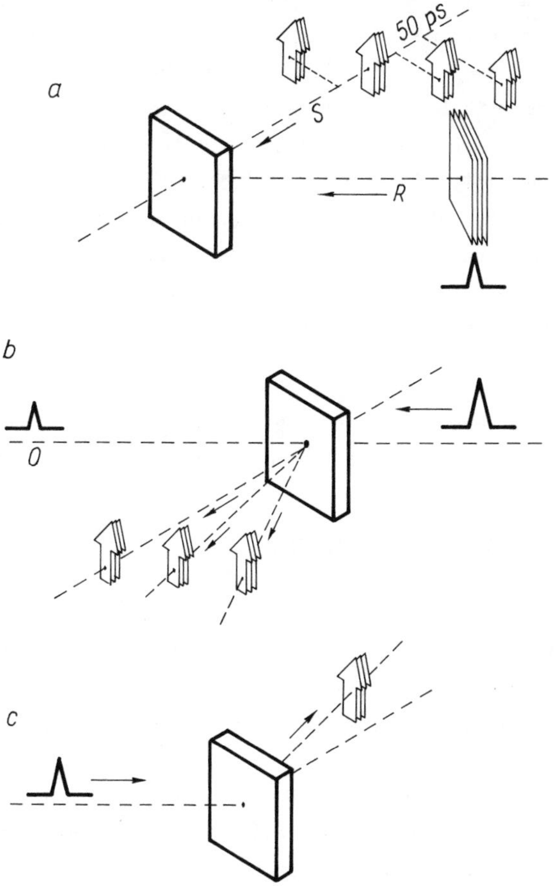

Figure 6: Recording (a) and read-out (b,c) of time-and-space-domain holograms. The holographed scene: sequenced picosecond flashes of light in the form of arrows from the left to the right; to the scene corresponds the irradiated field S. The reference pulse R, in dependence on its delay t_R, splits the signal into two parts: "future" and "past". In a thick hologram, the read-out towards the reference pulse (b) reconstructs a virtual image of the part of the signal corresponding to "future"; read-out in the opposite direction (c) recalls a wave, reversed in time as well as in space, from the part of the signal corresponding to "past". The schematic describes a really-performed experiment. Photos of the images reconstructed from holograms, with the numbers and shape of the arrows on them corresponding to the corollaries of the theory, are presented in [11,12,15, 23]

We can see that the spectral hologram $\bar{E}\bar{E}^*$ is burnt not only into the virtual part, ε, i.e. into the absorption spectrum, but also into the real part, i.e. into the refractive index. Thereby, the operator $\hat{H}$ expresses the Kramers-Kronig relation between these quantities. Consequently, along with the absorption ZPL, the zero-phonon "hooks" of the refractive index, always accompanying ZPL, appear with equal rights. It would not be correct to think as if the role of the spectral hologram in dispersion were reduced only to the ensuring of the response causality (see Figure 2) or as if the changes of the refractive index at each particular frequency in the region of holes were negligibly small as compared to the changes in absorption. Quite the contrary, in case of a δ-shaped signal, for instance, $\bar{E}\bar{E}^*$ has a frequency-dependent term of the form $\cos(\omega t_R)$, the Hilbert transform of which equals $\sin(-\omega t_R)$, i.e. the spectral holograms in absorption and dispersion are identical, being simply shifted with respect to each other along the frequency scale. If in this case in the maxima of one lattice the light passing through the plate is weakened by e^x times, then in the maxima of the other lattice, a shift of the wave phase by x radians takes place. It can also be demonstrated that on burning a single hole of the width $\sim 0.01\ \mathrm{cm}^{-1}$ in a plate with the optical density $D = 3$ up to total bleaching the group velocity at the hole frequency decreases so much that in a 1 cm-thick plate photons are delayed by $\sim 10^{-8}$ s. It is not excluded that such effect can find a spectroscopic application, for example, to increase the probability of the processes of nonresonant interaction of light with matter. By replacing (5) into the Maxwell equation it can be found how the read-out field is transformed on passing through the plate. In case of a thin hologram and weak field the effect of the hologram on the read-out field is manifested through the change of the complex transmittance.

$$\Delta K(\omega,\underline{r}) = (k/2)(1 + i\hat{H})\{\bar{E}\bar{E}^*\} \ . \tag{6}$$

The response of the hologram to the δ-shaped pulse is given by the reverse Fourier transform $\Delta K(\omega,\underline{r})$ (see Chap. 2). Thus, for two last terms in (3), which are of most interest, we obtain

$$\begin{aligned} E_S^{out}(\underline{r},t) \sim\ & Y(t - \underline{n}_R \circ \underline{r})\ S(\underline{r},t + t_R)\ \exp[i\omega_0(t + t_R - z/c)] \\ & + Y(t - \underline{n}_R \circ \underline{r})\ S^*(\underline{r},-t + t_R + \underline{n}_{2R} \circ \underline{r})\ \exp[i\omega_0(t - t_R - \underline{n}_{2R} \circ \underline{r})] \ . \end{aligned} \tag{7}$$

where $\underline{n}_{2R}$ is the unit vector, at an angle 2θ with the z-axis. On proceeding from (6) to (7) we resorted to the circumstance that in the temporal representation the convolution of the two frequency functions gives the product of the corresponding Fourier transforms.

We can see that the output field obeys the causality requirements: the Heaviside unit steps, Y, ensure the absence of the field at any point behind the hologram until the moment the front plane of the read-out pulse passes through this point.

Expression (7) describes two waves separated at some distance. In case the delay t_R is negative (on recording the reference pulse was incident on the plate prior to the signal), in the first term the function Y is superfluous. Consequently, in this case, the first term in (7) describes fully the reconstructed signal field, emerging from the hologram with the delay t_R with respect to the read-out pulse. The observer near the z-axis perceives the field as a virtual image of the recorded scene. It is essential to emphasize that a total reconstruction of the object scene in all its temporal dependence takes place, including the possible frequency modulation of the radiation of the sources. In case of positive t_R the cut-off of the signal pulse due to the function Y takes place. For example, if the signal consists of two pulses, the second of which reaches the plate after the reference pulse, then in the field restored in the direction z the first pulse is missing (see Figure 6b).

Finally, it is easy to get convinced that the second term in (7) describes the formation of the real image (the image is pseudoscopic, i.e. the depth of the scene has been "turned inside out") of the temporally reversed object scene in the direction $\underline{n}_{2R}$. Thereby, if t_R exceeds t_S, i.e. when the reference pulse is applied after the signal has passed through the plate, no cutting off the signal at the front plane of the read-out pulse occurs on reconstruction. On the contrary, in case of the uncut reconstruction of the virtual image this term turns to zero, i.e. the real image is absent.

Thus, in case of one direction of the recalling pulse either a virtual or a real reconstruction of the image (course of events) appears depending on the sequence of the delivery of pulses (reference and object) on recording.

Now a problem is to be discussed what novel results can be expected from the increase of the thickness of holograms. An analysis has shown [10] that in this case, a total reconstruction of the object scene in all its temporal dependence is possible, however, only in case of two directions of read-out pulses. Moreover, additional restrictions are put on the formation of the image, besides those resulting from causality: if the read-out pulse is applied in the direction of the referenece pulse, then only a virtual image is recalled; to reconstruct a real image (with the reverse time course) it is necessary to direct the read-out pulse antiparallel to the reference one - in this case, no virtual image appears (Figure 6b). Thus, it can be inferred that the increase of the thickness of a hologram results in the restriction of wave synchronism known in connection with thick holograms.

Experiments have fully confirmed the corollaries of the theory (see [11-15] and also Figure 1). The spatial resolution of holograms was checked by holographying an event by way of scattering a picosecond pulse from the surface of a 10-copeck coin. The field restored in this experiment excellently reconstructed the relief of the coin (see the photos of the holographic image in [11,12,15,23]).

The improved version of the theory [20] gives the following results:

1. Consideration of the non-vanishing width of the ZPL results in the appearance of a common multiplier $\exp(-2t/T_2)$ before response (7) (see Chap. 3).
2. The maximum diffraction efficiency (ratio of the energies of the stored and read-out pulses) for thin holograms is 7.3%, for thick ones, 8.1% (at optimal thickness of the plate); these values exceed considerably the analogous characteristics for ordinary absorbing holograms, as the time-and-space-domain hologram is always an amplitude-phase one.
3. In conditions of maximum diffraction efficiency the ratio of the energy of the recalled signal pulse to the energy of the read-out pulse, that passed directly (without diffraction) through the plate, may reach up to 54% for a thin hologram and 51% for a hologram of an optimal thickness.
4. Depending on exposure, additional responses due to the nonlinear character of the diminishing of the number of resonance molecules $g(\omega,\underline{r})$ propagate towards the direction of the read-out beam under the angles of $n\theta$ ($n = 2,3,\ldots$). These weak responses emerge with the delay nt_R with respect to the read-out pulse and represent (n-1)-fold autoconvolutions of the signal.

Let us regard the possibilities of holographying the field in the full meaning of the word: considering also the spatial and temporal dependence of the polarization of the field vector. It is known that in the media with photoinduced anisotropy it is possible to holograph the spatial dependence of polarization [24]. The dipole momenta of the electronic transition of impurity molecules in glassy matrices are oriented along all possible angles Ω. Therefore, a PHB by the light of certain polarization causes dichroism and birefringence in the hole region. These effects have found a spectroscopic application [25] and, evidently, they are also applicable in holography. Let us see how the above-presented theory must be modified to consider polarization and what are the conditions for a correct reconstruction of the vector field.

For this purpose two signals $S_x(\underline{r},t)$ and $S_y(\underline{r},t)$ must be introduced as the projections of the vector signal on the unit vectors $\underline{x},\underline{y}$. Such description, evidently, allows the representation of the field with any polarization (including elliptic) and its dependence on time at any point of the wave. For the reference pulse it is also necessary to introduce two complex components, the ratio between which determines either the orientation of the linear polarization, or the direction of the circular polarization. As, according to $(\underline{e}_\Omega \cdot \underline{E})^2$, the probability of the absorbing of the field $\underline{E}$ by the molecule depends on the vector of the orientation of its dipole moment of the transition, $\underline{e}_\Omega$, then after exposure the distribution of impurity centres acquires the orientational dependence $g_\Omega(\underline{r},\omega)$. Further, the read-out-field-induced polarization P in an impurity centre is anisotropic and proportional to the tensor product $[\underline{e}_\Omega \underline{e}_\Omega]$. Proceeding to the dielectric permittivity of the medium of the hologram, an averaging over all possible Ω must be performed, thereby, due to anisotropy in $g_\Omega(\underline{r},\omega)$, the permittivity remains a tensor. The transmittance of the plate will also be a tensor, for which we obtain (cf. (6)):

$$\Delta K(\omega,\underline{r}) = (3k/2)(1+i\hat{H})\{<[\underline{e}_\Omega \underline{e}_\Omega](\underline{\bar{E}}\cdot\underline{e}_\Omega)(\underline{\bar{E}}^*\cdot\underline{e}_\Omega)>_\Omega\} \ , \tag{8}$$

where $< >_\Omega$ describes averaging over orientations. The latter procedure breaks the tensor ΔK up into isotropic and anisotropic parts, by which it is possible to investigate the recording of the terms $\underline{\bar{E}\bar{E}^*}$ in case of various polarization of the read-out pulse. The results of the analysis are the following: (i) the primary signal wave is recalled without distortions in case of circular polarizations of the same directions of the field vectors of the reference and read-out pulses; the conjugated wave, in case of opposite directions of these polarizations; (ii) the signal wave is reconstructed with the polarization reflected around the unit vector $\underline{x}$ in case of (mutually) orthogonal linear polarizations of the reference and read-out pulses; (iii) in case of other geoemtries a mixing of the polarizational components of the signal takes place in the reconstructed field.

The first experiments confirm these conclusions [26].

5. Conclusion: Applications in Optical Data Processing

Owing to longevity and high efficiency of space-and-time-domain holograms, their applicational prospects are evident in solving the following tasks. For some of these tasks the first model experiments have been carried out:

1. Processing of picosecond signals: (a) algebraic summation (see [20]); (b) spatial-temporal convolution of signals by reading the hologram of one signal by the other signal; (c) filtration of spatial and temporal frequencies by transmitting the signal through the filter-hologram; (d) wavefront conjugation and time reversal of the signal by reading the conjugated wave from the hologram (see [11-13,15]).
2. A synthesis of picosecond optical signals through the scattering of pulses on a hologram constructed by means of PHB on exposure to a tunable laser or some other light source with appropriate spatial-spectral parameters.
3. Recognition of picosecond-domain events - the generalization of the holographic method of image recognition. If a light signal is delivered to the hologram, which coincides with one of the signals recorded on it earlier, and if the outcoming light is focussed, then a δ-shaped pulse appears at a definite point of the focal plane (see [11-15]).
4. Playback of an event by its fragment through the generation of a phantom event (see [27]) and constructing an associative memory.
5. Parallel information storage (and playback) into the spectral memory (see [27] and reviews on PHB in this issue).

The author is indebted to K.Rebane for useful counselling in compiling this review.

References

1 K.K. Rebane: Elementarnaya teoriya kolebatel'noj struktury spektrov primesnykh tsentrov kristallov (Nauka, Moskva 1968) [English transl.: Impurity Spectra of Solids (Plenum Press, New York 1970)]
2 K.K. Rebane: Zh. Prikl. Spektrosk. 37, 906 (1982) [English transl.: J. Appl. Spectrosc. 37, 1346 (1982)]
3 K.K. Rebane: in Proc. Conf. "Lasers 82", New Orleans 1982, p. 340
4 K. Rebane, P. Saari: J. Lumin. 16, 223 (1978); V.V.Hizhnyakov: in this issue
5 K.K. Rebane, L.A. Rebane: Pure Appl. Chem. 37, 161 (1974)
6 A.A. Gorokhovskii, R.K. Kaarli, L.A. Rebane: Pis'ma Zh. Eksp. Teor. Fiz. 20, 474 (1974) [English transl.: JETP Lett. 20, 216 (1974)]
7 A.K. Rebane, R.K. Kaarli, P.M. Saari: Pis'ma Zh. Eksp. Teor. Fiz. 38, 320 (1983) [English transl.: JETP Lett. 38, 383 (1983)]
8 A.K. Rebane, R.K. Kaarli, P.M. Saari: Opt. Spektrosk. 55, 405 (1983)[English transl.: Opt. Spectrosc. USSR 55, 238 (1983)]
9 A. Rebane, R. Kaarli, P. Saari, A. Anijalg, K. Timpmann: Opt. Commun. 47, 173 (1983)
10 P. Saari, A. Rebane: Eesti NSV Tead. Akad. Toim. Füüs. Matem. (Proc. Estonian SSR Acad. Sci.) 33, 322 (1984) (in Russian)
11 A. Rebane, R. Kaarli, P. Saari: Eesti NSV Tead. Akad. Toim. Füüs. Matem. (Proc. Estonian SSR Acad. Sci.) 34, 328 (1985) (in Russian)

12 A. Rebane, R. Kaarli, P. Saari: Eesti NSV Tead. Akad. Toim. Füüs. Matem. (Proc. Estonian SSR Acad. Sci.) 34, 444 (1985) (in Russian)
13 R.K. Kaarli, A.K. Rebane, K.K. Rebane, P.M. Saari: Izv. Akad. Nauk SSSR Ser. Fiz. 50, 1468 (1986)
14 P.M. Saari, R.K. Kaarli, A.K. Rebane: Kvantovaya Elektron. 12, 672 (1985) [English transl.: Sov. J. Quantum Electron. 15, 443 (1985)]
15 P. Saari, R. Kaarli, A. Rebane: J. Opt. Soc. Am. B3, 527 (1986)
16 T.W. Mossberg: Opt. Lett. 7, 77 (1982)
17 N.W. Carlson, L.J. Rothberg, A.G. Yodh, W.R. Babbitt, T.W. Mossberg: Opt. Lett. 8, 483 (1983)
18 N.W. Carlson, W.R. Babbitt, T.W. Mossberg: Opt. Lett. 8, 623 (1983)
19 P.M. Saari: In Trudy II Vsesojuznoj shkoly "PLAMYA", Vilnius, 1981 (Nauka, Moskva 1983) p. 199
20 R. Sarapuu, R. Kaarli: Eesti NSV Tead. Akad. Toim. Füüs. Matem.(Proc. Estonian SSR Acad. Sci.) 36, 299 (1987) (in Russian)
21 W.H. Hesselink, D.A. Wiersma: Phys. Rev. Lett. 43, 886 (1979)
22 Yu.N. Denisyuk: in Problemy opticheskoj golografii (Nauka, Leningrad 1981) p. 7
23 P.M. Saari: Izv. Akad. Nauk SSSR Ser. Fiz. 50, 751 (1986)
24 Sh.D. Kakichashvili, T.N. Kvinikhidze: Kvantovaya Elektron. 2, 1449 (1975) [English transl.: Sov. J. Quantum Electron. 5, 778 (1976)]
25 M. Romagnoli, M.D. Levenson, G.C. Bjorklund: J. Opt. Soc. Am. B1, 571 (1984)
26 A. Rebane, R. Kaarli: Eesti NSV Tead. Akad. Toim. Füüs. Matem. (Proc. Estonian SSR Acad. Sci.) 36, 208 (1987) (in Russian)
27 Ya.V. Kikas, R.K. Kaarli, A.K. Rebane: Opt. Spektrosk. 56, 387 (1984) [English transl.: Opt. Spectrosc. USSR 56, 238 (1984)]

Zero-Phonon Lines and Anomalous Photoelectric Properties of Ruby

A.A. Kaplyanskii, S.A. Basun, and S.P. Feofilov

A.F. Ioffe, Physico-Technical Institute, USSR Acad. Sci.
SU-194021 Leningrad, USSR

Abstract

A study into the spectral response of photoconductivity, $j(\nu)$, of the concentrated ruby Al_2O_3:0.4% Cr_2O_3 has revealed photocurrent resonances in the region of zero-phonon R-lines, ($^4A_2 \rightarrow {}^2E$),of Cr^{3+} absorption. On scanning the excitation frequency over the contour of R-lines in the external field $E_0 || C$ (C is the trigonal crystal axis), the photocurrent has been observed to *change the sign:* on exciting in the outer wings of R-line $j(\nu) > 0$, while in the inner wings the photocurrent flows opposite to the field, $j(\nu) < 0$ ("sign-alternating resonant photoconductivity"). The photocurrent in ruby is shown to be due to the hopping of the excess charge along the Cr^{3+} ions excited into the metastable ^2E state. The change of the sign of $j(\nu)$ originates from the frequency-dependent spatially-asymmetric excitation of the Cr^{3+} ions located close to the charge carrier (Cr^{2+} or Cr^{4+}), which is responsible for the actual charging (i.e. current) direction. The appearance of the spatial asymmetry under selective photoexcitation in zero-phonon lines of Cr^{3+} is due to a combined action of the "internal" Coulomb field of the charge carrier and the external electric field applied to the crystal on zero-phonon lines of Cr^{3+}. A micromechanism of the negative absolute photoconductivity and photoelectric instability of ruby observed under "nonresonant" excitation into the broad U-band is discussed in the light of the data obtained.

1. Introduction

One of the remarkable features of zero-phonon lines (ZPL) in the spectra of impurity centres is the high sensitivity of the line position to various influences on the energy levels of centres. This feature is observed and successfully applied in two types of spectroscopic investigations. In one of them the crystal is subjected to the action of a uniform external, either electric or magnetic, field or to a uniaxial stress. In these cases, ZPL in absorption (emission) spectra of the crystal exhibit splitting or shift due to the influence of external fields on the electronic levels of impurity centres. The analysis of the ZPL behaviour in external fields provides reliable information on the symmetry (degeneracy) of electronic levels and on the local symmetry of impurity centres. In other type of experiments, the inhomogeneous broadening of ZPL due to the effect of random fields, such as electric or magnetic fields or a stress induced by the defects (point, linear, etc.) always present in the lattice, is studied.

In this paper, we discuss a recently-observed effect [1], the nature of which is essentially connected with the simultaneous effect of both types of fields - an internal electric field creating the inhomogeneous broadening of ZPL, and a uniform external electric field producing a Stark splitting on the ZPL of impurity centres. The effect has been observed experimentally in ruby crystals ($Al_2O_3:Cr^{3+}$). The narrow resonance in the ZPL of Cr^{3+} ions has been observed in the photoconductivity spectrum of a concentrated ruby; on scanning the excitation frequency over the contour of ZPL the photocurrent changes its sign so that on exciting in a definite region of the ZPL contour the current flows opposite to the external electric field applied to the crystal ("sign-alternating resonant photoconductivity").

2. Some Spectroscopic and Photoelectric Properties of Ruby

The corundum crystal Al_2O_3 belongs to the centrosymmetric trigonal class D_{3d}. Al^{3+} ions occupy the positions in the lattice, which are of a polar trigonal site symmetry of C_3 point group. There are two such positions in the lattice (A and B) which are inversion-symmetric to each other. The Cr^{3+} ions, substituting Al^{3+}, may consequently occupy two inversion-symmetric polar positions A and B (Figure 1a). The Cr^{3+} ions have dipole moments which are equal in magnitude and parallel to the C axis of the crystal, their signs being opposite for ions in A and B positions.

In the optical absorption spectra of ruby, narrow ZPL, R_1 and R_2, corresponding

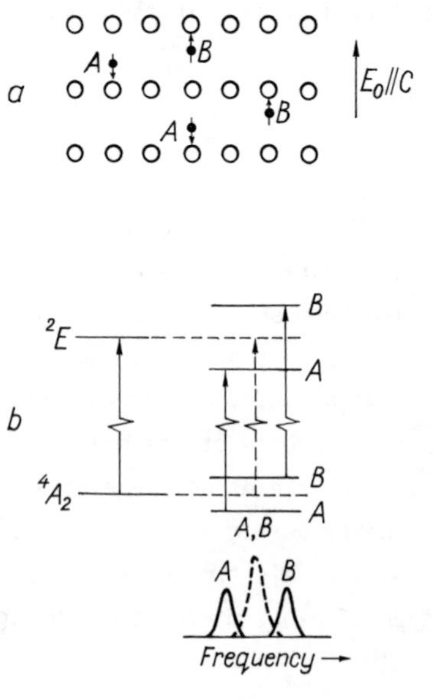

Figure 1: Polar positions of Cr^{3+} ions in Al_2O_3 lattice (a) and the pseudo-Stark splitting of spectral lines in the external field $E_o||C$ (b)

to the transitions from the quartet ground $^4A_2$ state to the $\bar{E}(^2E)$ (14418 cm^{-1}) and $2\bar{A}(^2E)$ (14447 cm^{-1}) levels belonging to the lowest excited ^2E state, are of the largest wavelength. The transitions from $^4A_2$ to the upper-lying quartet excited $^4T_2$ state produce an intense electron-vibrational U-band in absorption with the decay-broadened ZPL, $\nu_U = 16750$ cm^{-1}, and its vibrational replica ($\Delta\nu = 200$ cm^{-1}).

In the external electric field parallel to the trigonal axis ($E_0||C$), a linear doublet symmetric splitting of all ZPL takes place in the spectrum of ruby [2,3]. This, the so-called pseudo-Stark, splitting of ZPL is due to the linear Stark shifts (opposite in sign and equal in magnitude) of the energy levels of Cr^{3+} ions occupying A and B positions in the lattice (Figure 1 b). The form of the pseudo-Stark splitting does not depend on the sign of the electric field. The width of the doublet is $\Delta\nu = 2\Delta dE_0$, where Δd is the difference of the dipole moments of Cr^{3+} in the excited and ground states; in the case of R-lines $\Delta d = 0.39\cdot10^{-5}$ $cm^{-1}/(V/cm)$. The external field perpendicular to the trigonal axis ($E_0\perp C$) does not affect significantly the ZPL; no linear Stark effect is present.

At room and lower temperatures ruby is a perfect insulator. The Maxwell time of charge relaxation in ruby is of the order of years. It has been known before that some finite electric conductivity appears in ruby under illumination, but the photoelectric properties of ruby have practically been unexplored until recently.

Detailed studies of the photoelectric properties of ruby have been performed during the past few years on the so-called concentrated ruby with Cr concentration of some tenths of weight percent [4-9]. In the experiments performed at relatively low temperatures (77 K and lower), the argon laser excitation of crystals via the $^4A_2 \rightarrow {}^4T_2$ transition of Cr^{3+} has been accomplished. It has been found [6-9] that the character of the dependence of the photocurrent j on the applied field E_0 changes drastically with the orientation of the field E_0 in the crystal. In case of $E_0\perp C$, in the wide region of fields the current-voltage (i-V) characteristic $j^{\perp}(E_0)$ is of the common Ohmic type (Figure 2) [7]. However, with $E_0||C$ the i-V characteristic $j''(E_0)$ in the region of relatively small fields has a significant N-shaped anomaly: in a certain field interval, $-E_s < E_0 < +E_s$, the absolute conductivity (j''/E_0) is negative, i.e. the photocurrent flows opposite to the inducing electric field; in the $-E_t < E_0 < E_t$ interval, the differential conductivity ($\partial j''/\partial E_0$) is also negative. The values

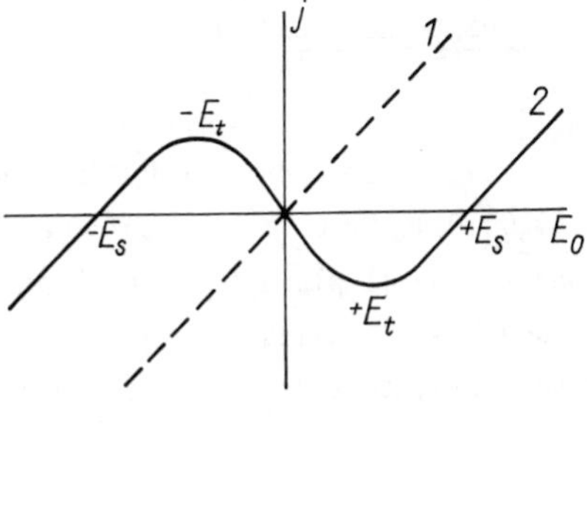

Figure 2: The schematic shape of the i-V characteristic of ruby in the field applied perpendicular (1) and parallel (2) to the trigonal axis

of $|E_s|$ are about 350 - 500 kV/cm, $|E_t| \simeq |E_s|/2$.

The existence of the N-shaped field dependence of the photocurrent $j''(E_0)$ results in the instability of the zero-field state of the crystal under optical pumping (with no external field applied): a spontaneous charge separation and the accompanying creation of domains of a uniform electric field, parallel to C-axis and equal in magnitude but opposite in sign ($\pm E_s$) in different domains, take place [5-9]. The formation of this photoinduced electric field is registered by the appearance of the pseudo-Stark splitting in the ruby spectrum after the optical excitation with no external field applied. Thus the value of the characteristic "critical" field E_s can be determined not only in the electric measurements of N-shaped i-V characteristic in electrically uniform samples (as the field in which the current j" changes its sign, see Figure 2) but also in spectroscopic measurements of the pseudo-Stark splitting of ZPL in the spectra of "nonuniform" samples, in which the photoinduced domain structure has been created. It has been found that the value of E_s depends critically on several conditions: temperature [5,7,8], concentration of chromium in the sample [4,10], light intensity [4,8].

At present the experimental picture of the anomalous photoelectric phenomena in ruby is rather complete and there exists also a phenomenological theory, which explains these phenomena on the basis of the N-shaped field dependence of the photocurrent $j''(E_0)$. However, the microscopic processes determining the macroscopic photoelectric properties of ruby, remain unclear. First of all, a question about the nature of photocurrent in ruby arises: an optical excitation with the quanta of visible light $h\nu \simeq 20000\ cm^{-1}$ seems to be able to induce only "internal" electron transitions in Cr^{3+} ions, not their photoionization (the charge transfer bands of $Al_2O_3:Cr^{3+}$ lie in the region $h\nu > 50000\ cm^{-1}$ [11]). On the whole, the type of charge carriers in ruby and the charge transfer mechanism are unknown. Neither the microscopic processes, causing the N-shape of i-V caharcteristic ($E_0 || C$) in the region of $j''(E_0)$ where the photocurrent flows opposite to the field, nor the nature of the critical dependence of this feature of the i-V characteristic on the changes of external conditions are known.

To answer these questions evidently the data on the spectral dependence of the photocurrent are significant. This dependence was studied in our experiments.

3. Experimental Observation of Sign-Alternating Resonant Photoconductivity in the ZPL Region

A schematic for the measurement of photocurrent excitation spectra in ruby is shown in Figure 3. Oriented crystal platelets of ruby Al_2O_3:0.4% Cr_2O_3 of the thickness $d \simeq 0.2$ mm, were used. The platelets were cut along the basal plane ($\perp C$) or along the plane containing the crystal axis ($||C$). The voltage U was applied to the transparent electrodes (SnO_2) deposited on the platelet surfaces, thus producing the elec-

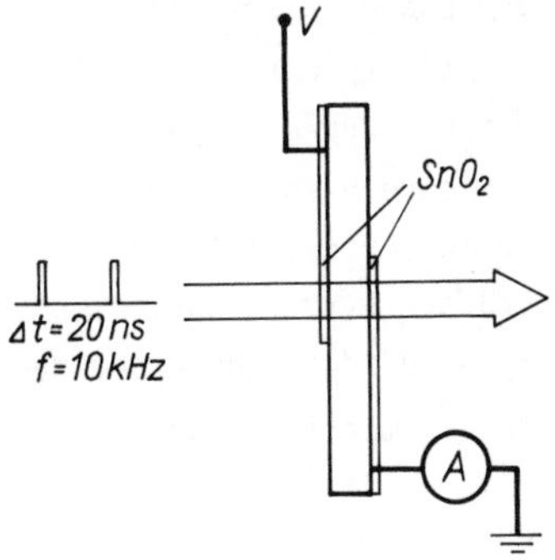

Figure 3: The scheme of the measurement of photocurrent excitation spectra

tric field $E_0 = U/d$ in the sample. The crystal was excited through the transparent electrodes by the radiation of a tunable dye laser and, with E_0 being fixed, the stationary photocurrent $j(\nu)$ was measured as a function of light frequency. A copper-vapour-laser-pumped dye laser was used; the exciting light pulses had the duration $\Delta t = 20$ ns, the repetition rate $f = 10$ kHz and the average power $P = 100$ mW. The frequency of the exciting light was changed in the interval from 14000 to 19000 cm^{-1} containing an R-line and a U-band of the ruby spectrum. Along with photoelectric measurements, a time-averaged luminescence spectrum of ruby in the R-line region, excited with the pulsed laser excitation of the sample, was registered. The experiments were carried out at low temperatures: 77 K and 4.2 K.

The experiments have shown that the photocurrent spectrum correlates with the absorption spectrum of Cr^{3+} ions in ruby, in particular, it exhibits clear resonances at the ZPL of ruby - at both R-lines and at the ZPL of U-band. The detailed shape of the photocurrent spectrum in the ZPL region depends on the direction of the field in the crystal, which is determined by the orientation of the crystal axis in the sample. At fixed optical excitation frequency the photocurrent depends linearly on the averaged excitation power (in the used pump intensity interval up to ≈100 W/cm^2).

In case $E_0 \perp C$ when there is no pseudo-Stark splitting of ZPL in the ruby spectrum, the i-V characteristic is linear (under nonresonant 514.5 nm argon laser excitation) [7], and photocurrent is always positive ($j^{\perp} > 0$). The photocurrent spectrum $j^{\perp}(\nu)$ reveals sharp maxima at R-lines (see Figure 4).

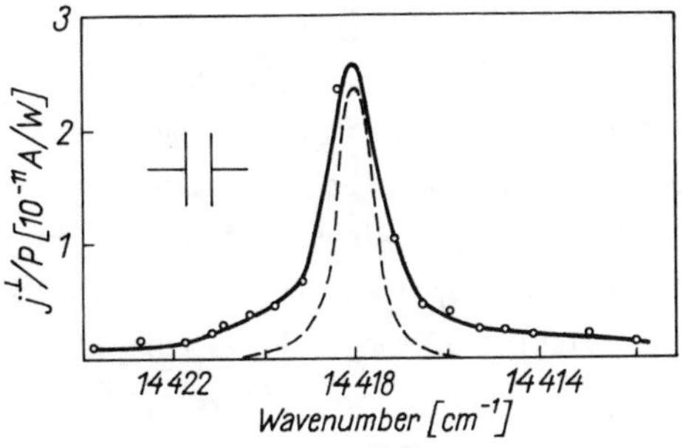

Figure 4: The resonant photocurrent excitation spectrum (solid curve) and the R_1 - line luminescence contour (dashed curve); $E_0 \perp C$, $E_0 = 420$ kV/cm

When the field is applied parallel to C-axis, $E_0||C$, a pseudo-Stark splitting of ZPL takes place in the ruby spectrum; the i-V characteristic (under 514.5 nm argon laser excitation) reveals a N-shaped peculiarity and the photocurrent spectrum $j''(\nu)$ depends significantly on the applied electric field E_0. Figure 5 shows the photocurrent spectra at the values of E_0 larger and smaller than the critical field E_s corresponding to the point of the sign change on the i-V characteristic (Figure 2) under nonresonant 514.5 nm argon laser excitation (in the samples used in the experiments $|E_s| = 475$ kV/cm). The contours of R-lines, exhibiting the symmetric pseudo-Stark splitting in the external field $E_0||C$, are also shown.

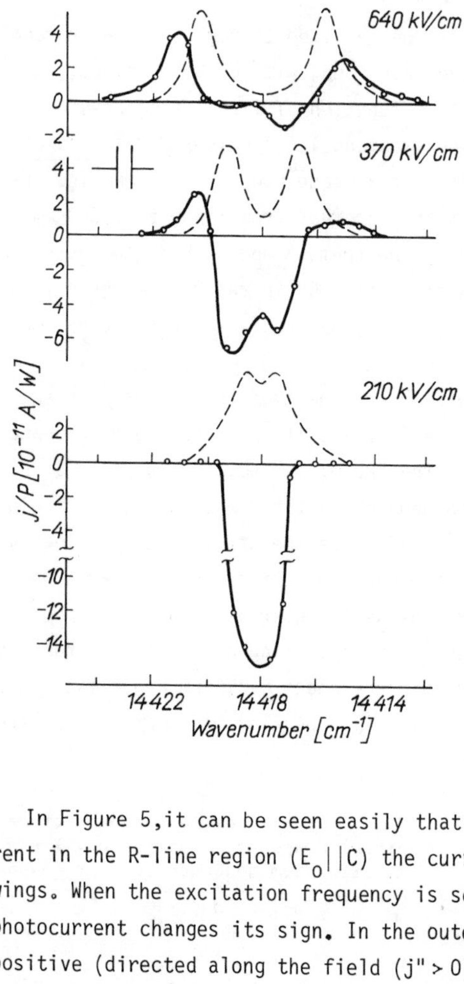

Figure 5: The photocurrent excitation spectrum in the R-line region (solid curve) and the R-doublet luminescence contour (dashed curve) at different values of the external field $E_0||C$

In Figure 5, it can be seen easily that under the resonant excitation of photocurrent in the R-line region ($E_0||C$) the current is preferentially excited in the line wings. When the excitation frequency is scanned over the split R-line contour, the photocurrent changes its sign. In the outer wings of R-doublet the photocurrent is positive (directed along the field ($j'' > 0$)). In the inner wings of R-doublet, it is negative ($j'' < 0$), i.e. it flows opposite to the field applied. The decrease of the external field from $E_0 > E_s$ results in the diminishing of the positive current in the outer wings (see Figure 5 a-c) and at $E_0 < E_s$ ($E_0 \approx E_s/2$) the positive current is not observed at all (Figure 5 c). At the same time, the negative current in the inner wings rises with the decrease of E_0 in this field interval and at $E_0 \simeq E_s/2$ the negative current becomes relatively strong.

Similar features in $j''(\nu)$ dependence are observed in the ZPL of the U-band of absorption (Figure 6) ($E_0 || C$). For the main structureless part of the U-band $j''(\nu)$ corresponds closely to the absorption coefficient $k(\nu)$ both in case $E_0 > E_s$ (photocurrent is positive, $j'' > 0$), and in case $E_0 < E_s$ (photocurrent is negative, $j'' < 0$). However, at the long-wavelength side of the U-band at the frequencies of the centres of the pseudo-Stark doublets of ZPL and its narrow vibrational replica sharp peaks of the negative photocurrent $j''(\nu)$ are observed. Thus, similar to the case of R-lines, negative photocurrent is excited in the spectral region inside the pseudo-Stark doublet of the U-band ZPL.

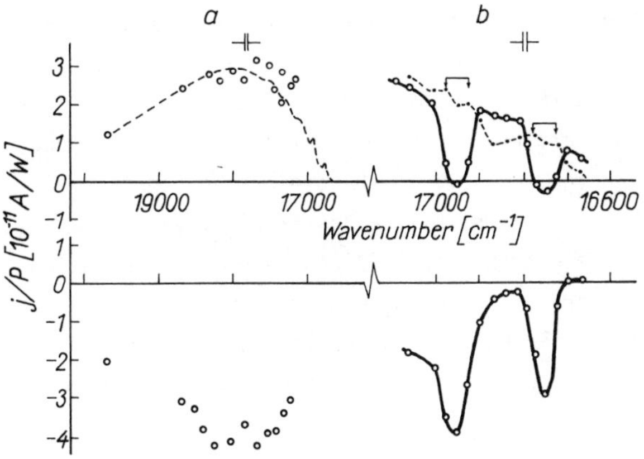

Figure 6: The photocurrent excitation spectrum in the regions of the broad structureless U-band (a) and U-band ZPL and its vibrational replica (b). The dashed curve is the absorption spectrum obtained from the excitation spectrum of R_1-luminescence

It should be mentioned that the magnitude of photocurrent and the main features of its spectral dependence practically do not differ at 77 and 4.2 K.

4. Sign-Alternating Resonant Photoconductivity and its Connection with the Spectrally-Selective Spatially-Asymmetric Excitation of Cr^{3+} Ions via the Inhomogeneously-Broadened ZPL Contour

The experimentally-observed correlation of photocurrent excitation spectra and Cr^{3+} absorption spectra provides direct evidence that photocurrent in ruby is produced through exciting Cr^{3+} to the metastable ^2E state. Such excitation may be performed "from below" - resonantly, by light absorption in R-lines ($^4A_2 \to {}^2E$) or "from above" - nonresonantly, via U-band ($^4A_2 \to {}^4T_2$), i.e. by the transition to $^4T_2$-state followed by nonradiative $^4T_2 \to {}^2E$ relaxation.

The photoconductivity under "internal" excitation of Cr^{3+} ions into ^2E state seems unexpected, because the ^2E energy (14400 cm^{-1}) is much less than Cr^{3+} ionization potential in Al_2O_3 lattice. The conclusion is that the charge transfer is accomplished along the system of Cr^{3+} impurity ions by photostimulated "jumps" of the "excess" electron or hole, corresponding to the anomalous (different from Cr^{3+}) charge state of the ions - Cr^{2+} or Cr^{4+} states, i.e. the current is produced by the directed charge transfer between the ions. The realization of the transfer of the electron (from Cr^{2+}) or the hole (from Cr^{4+}) to Cr^{3+} ion requires the excitation of this Cr^{3+} ion into ^2E-state. It is worth mentioning that the conclusion about the role of Cr^{2+} (Cr^{4+}) agrees with the well-known fact that Cr^{2+} and Cr^{4+} are always present in ruby in a small concentration or are produced by intense optical excitation [12].

Let us now explain the observed (Figure 5) sign-alternating spectrum of the photocurrent $j''(\nu)$ ($E_0||C$) under resonant excitation of photocurrent in the region of R-line. The main ideas of the interpretation of this unusual phenomenon are the following: (i) the electron (hole) jumps from the charge carrier Cr^{2+} (Cr^{4+}) to the Cr^{3+} ions situated near the charge carrier, (ii) the Cr^{3+} ions near the charge carrier are subjected to the strong Coulomb field of the carrier, which shifts the energy levels and resonant frequencies of the ZPL of "photoelectrically-active" Cr^{3+} ions (see Figure 7,a). Indeed, the charge carrier Cr^{4+} produces the electric field $\underline{E}_C = e\underline{r}/\varepsilon r^3$, where e is the electron charge, $\varepsilon = 11.3$ is the dielectric constant of Al_2O_3 and $\underline{r}$, the radius-vector. The z-component of the "internal" field $\underline{E}_C$ causes the linear Stark shift of the purely electronic transitions $^4A_2 - {}^2E$ in the neighbouring Cr^{3+} ions: $\Delta\nu = \Delta d|E_C|\cos\Theta$, where Θ is the angle between the direction from the charge carrier to the ion under consideration and the trigonal axis C, $\Delta d = 0.39 \cdot 10^{-5}$ $cm^{-1}/(V/cm)$ is the differential dipole moment characterizing the shift of the

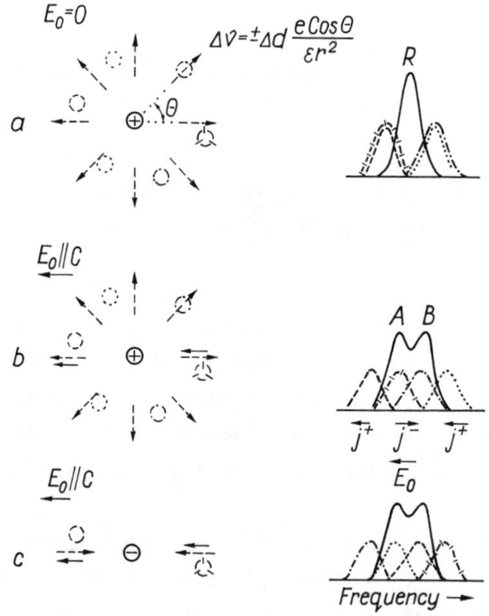

Figure 7: The scheme of the disposition of photoelectrically active Cr^{3+} ions near the positive (a,b) and negative (c) charges and the ZPL spectra of these ions (to the right). The different lines by which spectra and ions are shown mean polarity (A,B) and disposition ("to the right" and "to the left") of Cr^{3+} ions. Solid arrows - the external electric field, dashed arrows - the field of charge carriers. The C axis is horizontal

R-line in the electric field (see Chap. 2). For the characteristic distance $r = 2$ nm between Cr^{3+} ions in ruby with 0.4% chromium concentration the shift (with $\Theta = 0$) is $\Delta\nu = 1.2$ cm^{-1}, thus exceeding the R-line width. Thus, the ZPL frequencies of the "photoelectrically-active" Cr^{3+} ions situated in the vicinity of charge carriers and able (under optical excitation) to take part in the charge transfer process are shifted to the wings of the inhomogeneous R-line contour, corresponding to the main mass of Cr^{3+} ions in the lattice. It is essential that the direction of the "active" z-component of the charge carrier field E_C is of opposite sign for the ions situated "to the right" and "to the left" from the basic ($\perp C$) plane containing this charge. As a result of the doublet pseudo-Stark splitting in the internal field E_C the frequencies of R-transitions for A and B ions, situated "to the right", are disposed in the opposite wings of the "main" R-line; so are also the frequencies of the "left" A and B ions (for the latter the order of the distribution of the ZPL of A and B ions on the wings of the R-line is inverse). The corresponding distribution of the R-frequencies of the photoelectrically active Cr^{3+} ions, situated in the vicinity of the charge carrier, is schematically shown in Figure 7a.

Let us now consider the R-spectrum in the presence of the external electric field $E_0 || C$. This field induces the opposite shift of the frequencies of all A and B Cr^{3+} ions (pseudo-Stark splitting); the resulting distribution in the spectrum is schematically shown in Figure 7 b. The pseudo-Stark R-doublet of the main mass of Cr^{3+} ions and the frequencies of ZPL corresponding to the R-transitions in the photoelectrically active Cr^{3+} ions situated near positive charge carriers (Cr^{4+}), are depicted in the Figure. It is of significance that if the external field E_0 is present, the spatial regions "to the right" and "to the left" from the charge carriers become electrically nonequivalent, because in one of them (left) external (E_0) and internal (E_C^z) fields are added and in the other (right), they are subtracted. As a result, the pseudo-Stark R-doublet, formed by ions in the partially compensated field, is narrow and it is situated *inside* the main pseudo-Stark doublet. At the same time pseudo--Stark doublet, formed by ions in the strong summary field, is broad and is situated *outside* the main pseudo-Stark doublet. Thus, the photoelectrically active ions situated "to the right" and "to the left" from the charge carriers in the external field $E_0 || C$, differ in the spectrum.

Consequently, under the optical excitation of ruby in the external field $E_0 || C$ in the region of the pseudo-Stark R-doublet it is possible to perform the spatially selective excitation of the Cr^{3+} ions, situated in the vicinity of anomalously-charged chromium ions to a certain direction from these ions, to ^2E state.

It can be seen from Figure 7b that under optical excitation in the outer wings of the pseudo-Stark R-doublet, the Cr^{3+} ions situated *to the left* from the positive charge are preferentially excited to ^2E-state. In this case, the positively-charged hole can jump to the excited Cr^{3+} ion only on the left and thus the current flows *along* the external field E_0. Under excitation in the *inner wings* of the R-doublet the Cr^{3+} ions situated *to the right* from Cr^{4+} are excited to ^2E state and the hole can jump to the right, producing the photoelectric current opposite to the external field E_0.

Analogous conclusions about the mechanism of the sign-alternating resonant photoconductivity in the R-line are valid in the case of the negative sign of the charge carrier (Cr^{2+}) and the "electronic" character of the conductivity along the Cr^{3+} ion system. It can be seen from Figure 7c that here in the external field under the excitation in the outer wings the Cr^{3+} ions situated *to the right* from Cr^{2+} are excited to the ^2E-state;in this case the jump of the negative charge to the excited Cr^{3+} corresponds to the photocurrent along the electric field. The excitation in the internal wings excites the Cr^{3+} ions situated *to the left* from Cr^{2+} and the possible jump of the electron to Cr^{3+} (^2E) corresponds to the photocurrent opposite to the field. Consequently, the sign of photocurrent in the outside ($j'' > 0$) and inside ($j'' < 0$) wings of the R-doublet is the same in both cases, for Cr^{2+} and Cr^{4+}.

Thus, in the model under consideration the *direction* of photocurrent (the direction of charge transfer between Cr ions) is determined by the "geometrical" conditions, connected with the asymmetry (anisotropy) of the optical excitation of the photoelectrically active Cr^{3+} ions situated in the vicinity of charge carriers and actually capable of taking part in the charge transfer process. The external field E_0 in this model produces only the pseudo-Stark splitting of the R-line spectrum, including the spectrum of the Cr^{3+} ions "distorted" by the field of charge carriers. Thus, conditions for the spectrally-selective asymmetric excitation of photoelectrically active ions are created. The model explains qualitatively the main experimental features of the sign-alternating resonant photoconductivity of ruby excited in the zero-phonon R-lines in the field $E_0 || C$. First of all, the effective excitation of photocurrent preferentially in the wings of R-line can be explained. The change of the sign of the photocurrent on the scanning of the excitation frequency over the ZPL contour becomes also clear: the sign of the current is positive in the outside wings of R-line and negative in the centre of split ZPL. It can be seen from Figure 7b that under monochromatic excitation in the outside wings, corresponding to $j'' > 0$, preferentially only the ions of either A or B type are excited, opposite to the case of the excitation of the negative current in the central part of the doublet under which both types of ions are excited. This circumstance may be of significance in explaining the smaller amplitude of the positive current as compared to the negative one. Indeed, the charge jumps and the directed charge transfer are more facilitated in case both types of the ions (A and B) are excited as compared to the case when Cr^{3+} ions of only one (A or B) type are included.

Along with the mentioned "spectroscopic" effect of the external field on the asymmetry of spatial excitation, the effect of the field on the probability of the jumps of the charge to the neighbouring excited Cr^{3+} ion manifests itself in the dependence of the positive and negative photocurrents on the external electric field

$E_0 || C$ (Figure 5). Usually, in the case of hopping conductivity the probability of jumps in the direction of the field rises and the probability of the jumps opposite to the field decreases. This factor, obviously, must give its contribution to the effect of the increase of j^+ and the decrease of j^- observed at sufficiently large E_0 (Figure 5).

The effect of the external field on the probability of the jumps manifests itself more directly in the positive current under its resonant excitation via the R-lines in the field $E_0 \perp C$ (Figure 4). In this case, there is no pseudo-Stark splitting in the absorption spectrum (see Chap. 2) and in any point of the line contour the excitation of photoelectrically active Cr^{3+} ions is spatially symmetric (in Figure 7, if $E_0 \perp C$, it would be the symmetry of exciting "above" and "below" the charge carrier). The positive current $j^{\perp}$ can emerge only due to the difference between the probabilities of jumps "up" and "down", appearing in the external field. The maximum of $j^{\perp}(\nu)$ at the centre of the R-line is probably connected with the relatively large contribution of the jumps with a large projection on the $E_0 \perp C$ direction, for which the angle $\Theta = \pi/2$ and the corresponding frequency shifts of photoelectrically active ions from the centre of the line are small.

Thus, the external electric field plays a double role in the resonant sign-alternating photoconductivity of ruby ($E_0 || C$). On the one hand, the field E_0, via the splitting of the ZPL spectrum, creates conditions for the spectrally-dependent asymmetric excitation of the ions taking part in the charge transfer processes, thus determining the current direction. On the other hand, the E_0 field affects the probability of jumps in favour of jumps along the field, thus affecting the magnitude of the current.

5. Sign-Alternating Resonant Photoconductivity in ZPL and Photoelectrical Anomalies of Ruby

The elucidation of the nature of the experimentally-found sign-alternating resonant photoconductivity in the R-lines of ruby is essential for the understanding of all the other anomalous photoelectric properties of concentrated ruby. The most valuable results from this point of view are: (i) the determination of the mechanism of the photocurrent in ruby as a process of the transfer of the "excess" charge along the system of the metastable Cr^{3+} ions (^2E) and (ii) the establishing of the role of the microscopic spatial asymmetry of the optical excitation of Cr^{3+} ions as a factor determining the direction of the photocurrent in the external field and, in particular, the direction of the current opposite to the field - the most curious feature among the photoelectric properties of ruby.

The anomalous photoelectric properties of ruby (N-shaped i-V characteristic, photoelectric instability) were observed (see Chap. 2) under nonresonant Ar-laser pumping of ruby (mostly by 514.5 nm line) into the broad U and Y absorption bands. The results of the measurements of the spectral dependence of the photocurrent in the U-band region were discussed in Chap. 3 and shown in Figure 6. It is obvious that the narrow peaks of the negative photocurrent $j''(\nu)$ at the centres of the pseudo-Stark doublets of the ZPL of the U-band and its vibrational replica are of the same nature as the negative current $j''(\nu)$ in the inner region of the R-doublet. These peaks are due to the spatially-asymmetric excitation via $^4A_2$ of photoelectrically active Cr^{3+} ions, corresponding to the charge transfer opposite to the external field.

At the same time the nature of the photocurrent of different sign under nonresonant excitation via the structureless part of the U-band (Figure 6), in particular, the reason for the appearance of the negative current at $E_0 < E_s$, is more complicated. Due to the large width of the U-band, the spectral difference in the $^4T_2$ excitation of the Cr^{3+} ions situated near charge carriers is relatively small (the Stark shift of the U-band position, which is about 15 times larger than the shift of the R-line [3], is about 20 cm^{-1}, whereas the width of the U-band is about 2000 cm^{-1}). Thus, the spectral selectivity necessary for an efficinet spatially-asymmetric excitation of Cr^{3+} ions in the U-band, is suppressed and the pumping of the ^2E state of photoelectrically active Cr^{3+} ions via the broad U-band (with the subsequent $^4T_2 \rightarrow {}^2E$ relaxation) is spatially-symmetric. This is the principal difference between the case of a nonresonant excitation of the ^2E state via the U-band and the case of a resonant excitation of ^2E via the R-line. Under the spatially symmetric excitation of Cr^{3+} at any value of the field $E_0 || C$ the direction of the photocurrent must be positive due to the influence of E_0 on the probabilities of charge jumps along and opposite to the field. Nevertheless, the experiments show that the magnitude of the negative current under U-excitation, when there is no excitation asymmetry, is comparable to the magnitude of the current under the R-line excitation, when the asymmetry is nearly full.

The explanation of the negative direction of the photocurrent at $E_0 < E_s$ (and the N-shaped i-V characteristics in the case of $E_0 || C$) under *nonresonant* excitation of ruby via the broad U-band requires additional considerations. In principle, there are two possibilities of the explanation. Either we should hold that the nonresonant $^4T_2$ excitation of photoelectrically active ions to ^2E state is spatially symmetric and assume that in the field $E_0 < E_s$ the probabilities of charge jumps opposite to the field exceed the probabilities of the jumps along the field (in such case the polar symmetry of Cr^{3+} and Cr^{2+} (Cr^{4+}) positions may be significant, as it can lead to a strong linear-in-field shift of the energy levels of charge-transfer-active ions), or we should suggest that under nonresonant (U-band) excitation in the field $E_0 < E_s$, due to some *secondary* processes, an asymmetric excitation of photoelectrically active ions is realized to the benefit of the ions producing the negative current. Such secondary process could be (M.I.Dyakonov's idea) the excitation of active ions by the light of R-luminescence of the main mass of Cr^{3+} ions, excited via U-band. As it can be seen from Figure 7b, the character of the superimposition of the luminescence contour of the main R-doublet with the photocurrent excitation spectrum provides, in the case $E_0 < E_s$ (when R-doublet is not sufficiently resolved), the asymmetry of the spatial excitation to the benefit of the "right" active ions, thus producing the negative current. Unfortunately, the quantitative estimation and the results of specially performed experiments show that the intensity of R-luminescence is not sufficient (falling short by 2 orders of magnitude) to excite the negative current experimentally observed under U-excitation. Another possible secondary process is connected with the *nonradiative* transfer of ^2E excitation energy from the main mass of Cr^{3+} ions to active ions. This transfer (similar to the radiative one) can be more effective on the centre of the R-line contour, thus giving rise to the negative current. Additional experiments are necessary for the final clarification of the nature of the negative photocurrent in nonresonantly excited ruby[1].

In conclusion it should be mentioned that the phenomenon of sign-alternating resonant impurity-induced photoconductivity of ruby under excitation in ZPL region can be considered as a new "photoelectric" version of the selective laser spectroscopy of activated crystals inside the inhomogeneously-broadened ZPL contour. Indeed, the homogeneous broadening of ZPL by the internal electric fields of charge carriers is vitally needed for the phenomenon to exist. When the external field is applied, the action of the summary (external + internal) field, which separates the inhomogeneously-broadened ZPL of active ions in different positions relative to charged centres, must be taken into account. The latter circumstance immediately turns the spectrally-selective optical excitation into the spatially-selective excitation of ions, which in this case is registered through the charge transfer process resulting in the photocurrent of a definite sign.

The observation of charge transfer by selectively-excited Cr^{3+} ions shows the principal possibility of the observation of a selective hole burning in the inhomogeneously-broadened contour of the Cr^{3+} ions situated near Cr^{2+} or Cr^{4+} ions. Indeed, we have observed some decrease in the intensity of the resonant luminescence line selectively excited on the R-line wings. This observation gives indirect evidence about the decrease of the amount of Cr^{3+} ions with the corresponding ZPL frequency.

The authors gratefully appreciate the helpful discussions with M.I.Dyakonov and A.S.Furman.

---

1 In the recent studies [13] of the photocurrent kinetics, it has been shown directly that under nonresonant excitation the latter mechanism is responsible for the negative photocurrent, i.e. the resonant nonradiative ^2E excitation transfer from the main mass of ions to the neighbouring photoelectrically active ions.

References

1 S.A. Basun, A.A. Kaplyanskii, S.P. Feofilov: Pis'ma Zh. Eksp. Teor. Fiz. 43,344 (1986) [English transl.: JETP Lett. 43, 445 (1986)]
2 W. Kaiser, S. Sugano, D.L. Wood: Phys. Rev. Lett. 6, 605 (1961)
3 A.A. Kaplyanskii, V.N. Medvedev: Fiz. Tverd. Tela 9, 2704 (1967) [English transl.: Sov. Phys. - Solid State 9, 212 (1968)]
4 P.F. Liao, A.M. Glass, L.M. Humprey: Phys. Rev. B22, 2276 (1980)
5 S.A. Basun, A.A. Kaplyanskii, S.P. Feofilov: Pis'ma Zh. Eksp. Teor. Fiz. 37, 492 (1983) [English transl.: JETP Lett. 37, 586 (1983)]
6 M.I. Dyakonov: Pis'ma Zh. Eksp. Teor. Fiz. 39, 158 (1984) [English transl.: JETP Lett. 39, 185 (1984)]
7 S.A. Basun, A.A. Kaplyanskii, S.P. Feofilov, A.S. Furman: Pis'ma Zh. Eksp. Teor. Fiz. 39, 161 (1984) [English transl.: JETP Lett. 39, 189 (1984)]
8 S.A. Basun, A.A. Kaplyanskii, S.P. Feofilov: Zh. Eksp. Teor. Fiz. 87, 2047 (1984) [English transl.: Sov. Phys. - JETP 60, 1182 (1984)]
9 M.I. Dyakonov, A.S. Furman: Zh. Eksp. Teor. Fiz. 87, 2063 (1984) [English transl.: Sov. Phys. - JETP 60, 1191 (1984)]
10 S.A. Basun, A.A.Kaplyanskii, S.P. Feofilov: Fiz. Tverd. Tela 28, 929 (1986) [English transl.: Sov. Phys. - Solid State 28, 521 (1986)]
11 W.M. Fairbank, G.K. Klauminzer, A.L. Schawlow: Phys. Rev. B11, 60 (1975)
12 G.E. Arkhangelskii, Z.L. Morgenstern, V.B. Neustruev: Phys. Status Solidi 22, 289 (1967)
13 S.A. Basun, A.A. Kaplyanskii, S.P. Feofilov: Fiz. Tverd. Tela 29, 1284 (1987)

Zero-Phonon Lines in Superradiance

E.D. Trifonov

A.I. Herzen Pedagogical Institute,
SU-191186 Leningrad, USSR

Abstract

The general properties of cooperative spontaneous emission are discussed and a conclusion is drawn that an impurity crystal with intensive zero-phonon lines can be used as a source of superradiance. An analysis of the experiments on the amplification of spontaneous emission (close to superfluorescence) in impurity crystals is made. A threshold condition for superfluorescence is obtained. It gives the estimation of the possibility of superfluorescence in real systems. The advantages of the induced superradiance are discussed.

1. Introduction

Cooperative spontaneous emission of the inverted collection of identical atoms - superradiance - was predicted by DICKE in his original paper [1]. Subsequently superradiance was studied in detail in numerous theoretical works (one can find the bibliography in reviews [2,3]). This phenomenon is of interest because it offers an opportunity to generate ultrashort light pulses with duration less than the inverse width of the spectral line of ordinary luminescence. Indeed, the cooperative effect in spontaneous emission manifests itself more distinctly during the time interval less than the relaxation times T_2 and T_2^*, which determine the homogeneous and inhomogeneous widths of spectral lines, respectively. In this case the correlation of atomic dipole moments arises during the spontaneous emission and as a result the radiation intensity is proportional to the squared inversion density, which in its turn results in the decrease of the radiation decay time. This peculiarity differs superradiance from the stationary laser generation or amplified spontaneous emission, where radiation intensity is proportional to the first power of the inversion density. Since the large value of the phase memory time (or, in other words, the narrowness of the spectral line) is a favourable condition for superradiance to be observed, it is natural to expect that suitable objects to realize this effect are luminescent centres of a crystal which have the intensive zero-phonon lines.

2. Basic Features of Superradiance

The characteristic time scale τ_R of the superradiant pulse depends on the system's size. If the geometry of the system is such that Fresnel's number $F = D^2/\lambda L > 1$, where L is the sample's length D^2 is its cross section, and λ, the wavelength, then (see [4,5])

$$\tau_R = (3\gamma_0 n\lambda^2 L/4\pi)^{-1} . \tag{1}$$

Here γ_0 is the radiation constant of the spectral line, n is the initial density of inversion. Recall that $\gamma_0 = 32\pi^3 d^2/3\hbar\lambda^3$, where d is the transition matrix element of a dipole moment. According to (1) the time of cooperative spontaneous emission is $3n\lambda^2 L/4\pi$ times less than the radiative decay time γ_0^{-1}. This factor can be large enough. Note, however, that formula (1) is valid until $\tau_R > \omega_0^{-1}$, where ω_0 is the resonance frequency of the transition under consideration. The length L of the system capable of radiating in cooperative manner, is also limited. The maximum value of L, known as the cooperative length, is determined (by the order of magnitude) from the condition

$$L_c = c\tau_R , \tag{2}$$

where c is the light velocity in the medium. Using (1), one obtains

$$L_c = (3\gamma_0 n\lambda^2/4\pi c)^{-1/2} . \tag{3}$$

If the sample's length is less than the cooperative one, the superradiant pulse is a burst of radiation with an oscillation structure which occurs by the delay time $\tau_D = \tau_R\,[\ln(\pi\sqrt{N})]^2/2$ later than the inversion is created. The width τ_W of the first lobe of the radiation pulse is equal to $2\tau_R \ln(\pi\sqrt{N})$, where N is the number of the inverted atoms [4]. The superradiant delay can be a hundred times larger than the value of τ_R due to small fluctuations of the initial polarization (optical dipole moment) which the process of the cooperative radiation is induced by.

The results given above have been obtained for one-dimensional model of superradiance, which describes the case F = 1 best of all. Due to the angular divergency and the possible inhomogeneity of inversion over the cross section of the sample, the oscillation structure of the pulse can be smeared if F > 1. The conclusions, however, which have been drawn concerning the time scale, remain valid.

Obviously, the condition of phase memory conservation during superradiance can be expressed by inequality $T_2 > \tau_D$. Now let us regard the inhomogeneous broadening, which is characterized by the time T_2^*. The value T_2^{*-1} determines the inhomogeneous width of the spectral line. If the inequality $T_2^* > \tau_D$ is fulfilled, the influence of the inhomogeneous broadening upon superradiance is negligible. If the opposite inequality takes place, it seems at first sight that these atoms can participate in a cooperative process which belongs to the central region of the inhomogeneous distribution. In reality, since the delay time τ_D is proportional to τ_R and the latter is inversely proportional to the number of initially inverted atoms, the effective narrowing of the inhomogeneous distribution cannot alter the inequality $T_2^* < \tau_D$ cardinally. However, some tendency to this will be manifested due to the logarithmic factor in the expression of τ_D. Let us also remark that the several-order decrease of the number of inverted atoms, which is necessary for the appreciable alteration of the logarithmic factor, leads to the violation of the inequality $T_2 > \tau_D$. Thus, it is clear that the superradiant pulse can also be suppressed by the inhomogeneous broadening of the spectrum, though not so drastically as by the homogeneous one [6].

The Fridberg-Hartman relation $\alpha L = \bar{T}_2/\tau_R$, which connects the reduced time $\bar{T}_2 = (T_2^{-1} + T_2^{*-1})^{-1}$, the superradiant time τ_R, and the amplifier coefficient α [7,8], is extremely useful. In practice it is convenient to use this relation to determine τ_R since α and $\bar{T}_2$ are usually known. Thus, the inequality $\bar{T}_2 > \tau_R$ expresses the amplification condition and the inequality $\bar{T}_2 > \tau_D$, which is necessary for the pure superradiance to be observed, expresses that amplification must be strong enough. In principle, superradiance can also be observed at the spectrally-selective excitation. But in this case the pumping duration influences inevitably the superradiance dynamics.

3. The Attempts of the Observation of Superradiance in the Impurity Crystals

Till now superradiance has mainly been observed in gases [9]. At low pressure the relaxation times T_2 and T_2^* in these systems are of the order of 10-100 ns and the superradiant pulses are of nanosecond duration. The homogeneous width of zero-phonon lines in impurity crystals at low temperatures can approach the radiation one. But due to the considerable inhomogeneous broadening ($T_2^* = 10\text{-}100$ ps) "pure" superradiance can be observed only in the picosecond region.

Recently some experiments were performed in which the coherent amplification of spontaneous emission was observed on zero-phonon lines. Let us analyse these experiments to ascertain to what extent the observed radiation can be interpreted as superradiance.

In very interesting experiments by FLORIAN et al. [10] KCl crystals with O_2^- centres were excited by 30 ps fourth harmonic pulses of a mode-locked Nd:YAG laser (266 nm). The spectrum of this system, which contains a series of zero-phonon lines, has been investigated by K.K.REBANE and L.A.REBANE [11]. At sufficiently low temperature ($T < 30$ K) and at the excitation power above the threshold value of 20 GW/cm^2 the amplified spontaneous emission was observed on zero-phonon lines of the luminescence spectrum, which corresponds to the vibronic transitions 0-10, 0-11 (with the wavelengths of 592.8 nm and 629.1 nm). The intensity of this emission exceeded that of the ordinary spontaneous emission by 10^4 times. The measured delay time of the pulses varied from 0.5 to 6 ns and the mean width of the spectrum turned out to be two times less than that of the fluorescent spectrum (0.042 nm at 4.2 K). The excitation pulses were focused to a spot with the diameter of about 0.1 mm. So for the length of the sample about 1 cm the Fresnel number was nearly 1. The initial inversion density was $n \simeq 10^{17}$ cm^{-3}, which might be somewhat overstated. The inhomogeneous relaxation time can be estimated by the width of the spectral line: $T_2^* \simeq 30$ ps. The homogeneous relaxation time is of the same order: $T_2 \simeq 100$ ps. The radiation constant of a zero-phonon line must be determined, taking into account the Franck-Condon factor. According to our rough estimation it is about 10^{-3} of the whole radiation constant of the allowed transitions $\gamma \simeq 10^8$ s^{-1}. Thus $\gamma_0 \simeq 10^5$ s^{-1}. The data given above make it possible to determine the characteristic time of the superradiance $\tau_R \simeq 10^{-13}$ s. At a smaller value of the initial inversion density ($n \simeq 10^{16}$ cm^{-3}) one has $\tau_R \simeq 10^{-12}$ s. Then for the delay time of the superradiant pulse we obtain $\tau_D \simeq 100\ \tau_R = 0.1$ ns (here we put $N = 10^{12}$), in this case the gain is equal to $\alpha L = T_2^*/\tau_R \simeq 30$. It is easy to see that at such gain the criterion of pure superradiance is violated. That is why the intensity of the superradiant pulse can be suppressed appreciably. This also explains the narrowing of the inhomogeneous spectrum. We have no opportunity to dwell upon more exquisite peculiarities of this experiment. Some of them have been explained in paper [12].

The other pretender to superradiance is the pure electron transition ${}^1B_{2u} - {}^1A_g$ of a pyrene molecule injected into a diphenyl crystal. The amplification of radiation on this transition has been investigated by ZINOV'EV et al. [13]. The experiment was carried out at 4.2 K. The pumping was fulfilled by a 10 ns pulse of the third harmonic of a Nd:YAG laser. On reaching the threshold excitation power of $5 \cdot 10^5$ W/cm^2 the time of radiation was reduced to 5-6 ns (compared with the natural lifetime of 110 ns). The width of the considered zero-phonon line is conditioned by the inhomogeneous broadening $T_2^* = 100$ ps. The sample's geometry is such that the Fresnel number is much more than 1. In principle, superradiance is possible in such system, the favourable condition is the small homogeneous broadening, $T_2 = 10^{-7}$ s. However, for the value of $\tau_R = 10^{-10}$-10^{-9} s, which is given in [12], even the gain condition is not fulfilled, $\alpha L = T_2^*/\tau_R < 1$. Moreover, the considerable duration of the pumping must exercise significant influence on the superradiant pulse dynamics.

The experiments, which were described above, enable us to expect that zero-phonon lines in impurity crystals may really serve as sources of superradiance, but to obtain the pure effect it is necessary to overcome some difficulties connected with the influence of inhomogeneous broadening and with the realization of the sufficiently short pulse pumping.

4. Induced Superradiance

Till now we considered the cases where superradiance of the inverted system was initiated by the fluctuations of atomic polarization. The value of these fluctuations per unit atom is of the order of $1/\sqrt{N}$, where N is the number of initially inverted atoms in the system. This small parameter determines the delay time of the superradiant pulse: $\tau_D = (\tau_R/2)\,[\ln(\pi\sqrt{N})]^2$. Such type of the cooperative emission is commonly called superfluorescence. As it was shown in Chap.2, the conditions $T_2, T_2^* > > \tau_D$ have to be fulfilled to observe pure superfluorescence. Moreover, the duration of the excitation must be less than the delay time τ_D. These rigid restrictions are withdrawn for induced superradiance, the idea of which is the following. The coherent resonant pulse of the area $\Theta = (d/\hbar)\int_0^\infty \varepsilon(t)dt$, where $\varepsilon(t)$ is the amplitude of the electric field, is injected into an inverted system, for which the amplification condition $\alpha L = (\bar{T}_2/\tau_R) > 1$ is ensured. The optical atomic polarization, induced by this pulse during its propagation through the system, is proportional to Θ. It can considerably exceed the fluctuation value $1/\sqrt{N}$ leading to the decrease of the delay time. It is desirable that the duration of the injected pulse should be less than T_2, T_2^* and that it should be of the same order of magnitude as τ_R.

To realize the induced superradiance the gain required is not so high as the one needed for superfluorescence. On the contrary, the gain for which superfluorescence is practically absent is optimal for induced superradiance. This will permit the creation of the inverted system by a sufficiently long pumping which does not affect the output pulse dynamics.

Let us find the threshold gain $(\alpha L)_0$, for which the intensity of superfluorescence (into some space angle) is equal to that of the ordinary spontaneous emission (into the whole space angle). First we will consider the system with the Fresnel number equal to 1. In this case, the superfluorescence space angle is equal to the diffraction one. To estimate $(\alpha L)_0$ we will restrict ourselves to the time interval, during which the inversion is almost constant. In this approximation, the number of cooperatively emitted photons per unit time interval is equal to (see [12])

$$I(t) = (2\tau_R)^{-1}\,[1 + \int_0^t t'^{-1} I_1(\sqrt{2t'/\tau_R})\exp(-2t'/\bar{T}_2)dt'] \quad , \tag{4}$$

where I_1 is the modified Bessel function of the first order. Here the inhomogeneously broadened line is supposed to be Lorentzian. Note that in the limit of $(\alpha L)\to 0$ it follows from (4) that $I = (2\tau_R)^{-1}$, which coincides with the intensity of the ordinary spontaneous emission into the diffraction space angle. We obtain the stationary value of cooperative emission intensity in the linear regime (that is with the constant inversion) putting $t\to\infty$ in (4) (see [15])

$$I_{st} = (\alpha L/2\bar{T}_2)\,[1 + (\alpha L/4)\,{}_2F_2(3/2, 1; 2, 3; \alpha L)] \quad , \tag{5}$$

where ${}_2F_2$ is the degenerate hypergeometric function. The value of $I_{st}\bar{T}_2$ as a function of the gain αL is shown in Figure 1.

The threshold gain $(\alpha L)_0$ can be found from the condition

$$I_{st} = \gamma_0 N \quad . \tag{6}$$

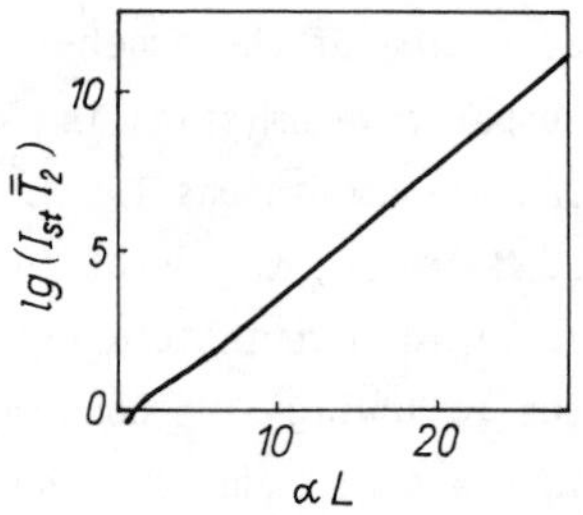

Figure 1: The intensity of the cooperative spontaneous emission in linear regime for the system with Fresnel number F = 1

For example, by using the experimental data of [10], we obtain the threshold value of $(\alpha L)_0 \simeq 28$.

If the Fresnel number of the inverted sample exceeds 1, condition (6) is to be changed by the following one:

$$F^2 I_{st} = \gamma_0 N \quad . \tag{7}$$

Let us use (7) to determine $(\alpha L)_0$ for the experiment [13]. On the strength of the data $N = 10^{16}$, $\bar{T}_2 = 10^{-10}$ s, $\gamma_0 = 10^7$ s^{-1}, $F = 10^4$ we obtain $(\alpha L)_0 = 13$.

Certainly, in the underthreshold regime the amplified radiation always exceeds the ordinary spontaneous emission at the same space angle, but the evolution of inversion is determined not by superfluorescence, but by the radiation decay or by other damping processes. That is why the inversion in such systems can be created by a comparatively slow pumping.

Note that the underthreshold regime is not necessary for the realization of induced superradiance. Originally it has been observed just for the condition, for which superfluorescence is also possible. We are referring to very interesting experiments by VREHEN and SCHUURMANS [16] in cesium vapours. The purpose of these experiments was to determine the injected pulse area Θ_0, for which the delay time of induced superradiance is equal to the delay time of superfluorescence. The value of Θ_0 was found to be $2/\sqrt{N}$ as it had been predicted by the theory.

The induced superradiance was observed by VARNAVSKII et al. [17,18] in crystals of Nd:YAG ($\lambda = 1.06$ µm, $\gamma_0^{-1} = 0.25$ ms) and ruby ($\lambda = 0.69$ µm, $\gamma_0^{-1} = 3$ ms). At liquid nitrogen temperature the relaxation times were $T_2 = 100$ ps, $T_2^* = 20$ ps for Nd:YAG and $T_2 = 100$ ps, $T_2^* = 60$ ps for ruby. By using the measured values of $\alpha L = 3$-5 and the relaxation times, one can determine the superradiance time τ_R: $\sim$10 ps for ruby and $\sim$7 ps for garnet. In its turn it enables us to estimate the initial inversion density, which is equal to 10^{16} cm^{-3} for ruby and 10^{15} cm^{-3} for garnet. The Fresnel number of the samples investigated was about 10^3. With the aid of these data we find the threshold value of the gain to be about 10 in both cases. This is in agreement with the experimental ability to create the inversion by a long pumping. The sources of input pulses were respectively the mode-locked Nd laser on silicate glass and the mode-locked ruby laser. When the area of input pulses was equal to $\Theta \approx 10^{-1}$ and their duration was 10-20 ps, two fragments could be distinguished on the densitograms. The first was the amplified input pulse and the second, the superradiant pulse, which followed with a certain delay.

The superradiant pulse is caused by the evolution of the optical polarization which is induced by the input pulse. In this experiment, the superradiant pulse was relaxed because the times T_2^* and τ_R are of the same order of magnitude and so the gain αL was not large enough.

To sum up we may note that zero-phonon lines in impurity crystals are objects with extremely good prospects for the realization of superradiance. First experiments in this field have already given important results. The induced superradiance is of special interest. It does not place so rigid demands upon the observation conditions and has a number of advantages such as the absence of the cooperative length limit, the possibility of using prolonged pumping and of controlling the pulse starting moment.

References

1 R.H. Dicke: Phys. Rev. 93, 99 (1954)
2 M. Gross, S. Haroche: Phys. Rep. 93, 301 (1982)
3 A.V. Andreev, V.I. Emel'yanov, Yu.A. Ilinskii: Usp. Fiz. Nauk 131, 653 (1980) [English transl.: Sov. Phys. - Usp. 23, 493 (1980)]
4 J.C. MacGillivray, M.S. Feld: Phys. Rev. A14, 1169 (1976); A23, 1334 (1981)
5 E.D. Trifonov, A.I. Zaitsev, R.F. Malikov: Zh. Eksp. Teor. Fiz. 76, 65 (1979)
6 R.F. Malikov, V.A.Malyshev, E.D.Trifonov: in Teoriya kooperativnykh kogerentnykh protsessov v izluchenii, ed. by E. Trifonov (Pedagogical Institute, Leningrad 1980) p. 3
7 R. Freidberg, S.R. Hartmann: Phys. Rev. A13, 495 (1976)
8 R.F. Malikov, V.A. Malyshev, E.D. Trifonov: Opt. Spektrosk. 53, 652 (1982) [English transl.: Opt. Spectrosc. USSR 53, 387 (1982)]
9 Q.H.F. Vrehen, H.M. Gibbs: in Dissipative Systems in Quantum Optics, Topics in Current Physics, Vol. 27, ed. by R. Bonifacio (Springer, Berlin 1982) p. 111
10 R. Florian, L.O. Schwan, D. Schmid: Phys. Rev. A29, 2709 (1984)
11 K.K. Rebane, L.A. Rebane: Pure Appl. Chem. 37, 161 (1974)
12 F. Haake, R. Reibold: Phys. Rev. A29, 3208 (1984)
13 P.V. Zinov'ev, S.V. Lopina, Yu.V. Naboikin, N.B. Silaeva, V.V. Samartsev, Yu.E. Sheibut: Zh. Eksp. Teor. Fiz. 85, 1945 (1983) [English transl.: Sov. Phys. - JETP 58, 1129 (1983)]
14 R.F. Malikov, E.D. Trifonov: Opt. Commun. 52, 74 (1984)
15 M.G. Benedikt, E.D. Trifonov: Opt. Spektrosk. 61, 681 (1986)
16 Q.H.F. Vrehen, M.F. Schuurmans: Phys. Rev. Lett. 42, 224 (1979)
17 O.P. Varnavskii, A.N. Kirkin, A.M. Leontovich, R.F. Malikov, A.M. Mozharovskii, E.D. Trifonov: Zh. Eksp. Teor. Fiz. 86, 1227 (1984) [English transl.: Sov. Phys. - JETP 59, 716 (1984)]
18 O.P. Varnavskii, V.V. Golovlev, A.N. Kirkin, R.F. Malikov, A.M. Mozharovskii, M.G.Benedikt, E.D. Trifonov: Zh. Eksp. Teor. Fiz. 90, 1596 (1986) [English transl.: Sov. Phys. - JETP 63, 937 (1986)]

Transient Phenomena in Mössbauer Gamma Radiation

E. Realo

Institute of Physics, Estonian SSR Acad. Sci.
SU-202400 Tartu, Estonian SSR, USSR

Abstract

Transient effects observed in Mössbauer coincidences and in the stepwise phase modulation of the Mössbauer radiation are considered both theoretically and experimentally. The main time-frequency dependences of the transmission and scattering transients are illustrated by corresponding experimental results on ^{57}Fe and ^{119}Sn Mössbauer resonances.

1. Introduction

In Mössbauer spectroscopy, the resonance on zero-phonon gamma lines with nearly natural linewidths has been accomplished. In this case the emission and absorption linewidths Γ_0 are determined by the lifetime τ_0 of the corresponding excited nuclear level $\Gamma_0 = \hbar/\tau_0$. It follows from this circumstance that the time and frequency dependences of the Mössbauer radiation are correlated and that the investigations performed in time and frequency domains are basically equivalent. At the same time a number of examples can be found in practical Mössbauer spectroscopy, which confirm that the most comprehensive picture of resonant interactions can be obtained from a complete time-frequency dependence of the radiation intensity transmitted through or scattered from a resonant absorber. Therefore, it is important to study the fundamental features of the time-dependent interactions of resonant gamma radiation with the matter containing resonant nuclei.

Time-frequency measurements assume the presence of a well-defined point on the time scale (a time reference) relative to which time dependences of intensity or time-resolved spectra (Mössbauer spectrochronograms - MSCG) are recorded. For the Mössbauer radiation a natural internal time reference is the moment t_0, when the excited resonant nuclear level is formed in a gamma source. Depending on the particular decay scheme of a resonant isotope used, a signal strongly connected to t_0 is fixed as the moment of the detection of the gamma quantum preceding the Mössbauer transition (e.g. ^{57}Fe) or the X-ray quantum accompanying this transition (^{119}Sn). As a matter of fact, this method of time-dependent investigations - Mössbauer coincidences [1] - ensures the time count starting from the instantaneous radiation switch-in. Another type of time reference is employed in the studies of the Mössbauer radiation under an external periodic high-frequency modulation [2,3].

In this case the time-dependent radiation intensity is recorded relative to a fixed phase value of the exciting modulation signal - an external reference point. In the case of a stepwise Doppler modulation the moment T, when a stepwise violation of stationary resonance conditions occurs, serves as a time reference. This situation may be interpreted as the phase-switching of the gamma radiation.

A resonant absorber or a scatterer responds to the switch-in or to the phase-switching of the incident radiation and complicated time-frequency features - transient phenomena (TPh) - can be observed in the transmitted or scattered intensity. Depending on the type of time reference in the present consideration, TPh are distinguished in the case of switch-in or phase-switching of the resonant gamma radiation. From the established analogy between zero-phonon lines (ZPL) in the Mössbauer effect and in optics it follows [4-6] that the results on TPh of resonant zero-phonon gamma radiation are of interest for other fields of spectroscopy as well. The studies of TPh in solids by Mössbauer spectroscopy enable one to elucidate various relaxation processes. Actually the characteristic times of electron spin-spin, spin-lattice and cross relaxation, superparamagnetism and post-decay charge relaxation fall into the time region accessible for Mössbauer spectroscopy (10^{-11}-10^{-6} s). Although in some experiments on ^{57}Fe isotope such relaxations have been found [7-12] and explained theoretically [13-16], a number of problems concerning TPh in the case of various relaxation processes need further development.

The switch-in transients in Mössbauer spectroscopy were found already in 1960 [1] and a number of investigations have been performed in this field. Phase-switching transients of the Mössbauer radiation were realized and interpreted recently [17-18], therefore the prospects and possibilities of their application are still to be clarified. At the same time there is no doubt that these works have laid basis for a new method of time-frequency studies in Mössbauer spectroscopy.

TPh are well known in laser spectroscopy [19-23] and NMR [24-26] where they are used to measure various relaxation times. The free-induction decay, observable on using weak radiofrequency or laser fields, is a close analog of the phase-switching transients of the Mössbauer radiation. The corresponding strong field effects (spin and photon echoes or nutations), connected with the saturation of the polarization in the sample, are not observable in Mössbauer spectroscopy due to low intensities of the available gamma sources. Another essential distinction between the TPh of optical and Mössbauer radiation consists in differences in the broadening of ZPL and in different possibilities of detecting various kinds of secondary emission. While in laser spectroscopy, besides ZPL, TPh can be observed also on phonon wings, in scattering and luminescence (including hot luminescence) lines, in Mössbauer spectroscopy TPh can be detected exclusively in the ZPL of transmission and scattering. At the same time an extremely small line broadening in comparison with the natural width is a great advantage of resonant gamma radiation. In optics, ZPL are usually strongly broadened and natural linewidths cannot be realized. The theory of TPh for various types of secondary emission in the case of the switch-in of optical radiation has been developed in [27-29]. General conclusions drawn from the theory of ZPL are in good qualitative accordance with the theory and experiments on the TPh of Mössbauer lines.

In the present paper, we restrict ourselves to the consideration of general properties of TPh in transmission and scattering of the Mössbauer gamma radiation for both switching types. The examples of experimental work on TPh in the isotopes ^{57}Sn and ^{119}Sn are taken from authors' papers.

2. Switch-in Transients

2.1. Transmission

The TPh of the Mössbauer gamma radiation are well described in terms of the classical theory. The consideration usually assumes limited radiation intensities (saturation effects are neglected), an extremely sharp resonance and a near-unity value of the refractive index. A single-line gamma source and a resonant absorber are assumed to consist of damped harmonic oscillators with the central frequencies ω_s, ω_a and the half-widths Γ_s, Γ_a, respectively. At random time moments, t_o, the thin Mössbauer source emits a field amplitude of gamma radiation, which can be described by the following expressions [1]:

$$E_s(t - t_o) \sim \exp[-i\omega_s(t - t_o) - \Gamma_s(t - t_o)/2]\ \Theta(t - t_o) \tag{1}$$

or

$$E_s(\omega) \sim \frac{1}{2\pi_i} \int_{-\infty}^{\infty} \frac{\exp[i\omega(t - t_o)]\, d\omega}{\omega - \omega_s - i\Gamma_s/2} = \int_{-\infty}^{\infty} \varepsilon(\omega) \exp[i\omega(t - t_o)]\, d\omega \quad , \tag{2}$$

where the step function $\Theta(t - t_o) = 1$, when $t \geq t_o$ and is equal to zero, when $t < t_o$. When the described amplitude is incident on the resonant absorber, each monochromatic component $\varepsilon(\omega) \exp[i\omega(t - t_o)]$ excites forced oscillations in its volume. The absorber acts as a medium with a complex refractive index, which modifies the amplitudes and phases of the individual frequency components transmitted through the absorber. The absorber response to the external excitation at the frequency ω_s is given by [1]:

$$R_a(\omega) = \exp[-ib/(\omega - \omega_a - i\Gamma_a/2)] \quad , \tag{3}$$

where b takes into account the resonant thickness of the absorber, $b = n\sigma_o f_a \Gamma_o/4$, n is the surface density of resonant nuclei, σ_o is the resonance cross-section, f_a is the Debye-Waller factor of the absorber. For a thin absorber ($b < \Gamma_o$) the thickness parameter b causes the line broadening from Γ_a to $\Gamma_a + b$ (thickness broadening). In general, $\Gamma_s = \Gamma_o + \gamma_s$ and $\Gamma_a = \Gamma_o + \gamma_a$ take into account the homogeneous or inhomogeneous broadening of the source and absorber lines, γ_s and γ_a, respectively, when the latter can be described by single exponential decay constants. Here is a possibility of getting information about relaxation processes or distributions of internal fields in solids. An analysis of TPh [18] has shown that the individual contributions of γ_s, γ_a and b to the total linewidth can be separated from a single experiment.

As a result of passing the radiation through the absorber the amplitude $\varepsilon(\omega)$ is replaced by $\varepsilon(\omega)\, R_a(\omega)$. The dependence of the transmitted field amplitude on the time t ($t_o = 0$) and the frequency shift of line centres $\Delta\omega = \omega_s - \omega_a$ can be calculated by integrating in the frequency domain:

$$E(t,\Delta\omega) = \int_{-\infty}^{\infty} \varepsilon(\omega)\, R_a(\omega)\, e^{i\omega t}\, d\omega \quad . \tag{4}$$

The transmitted intensity, as usual, equals to the square of the field amplitude:

$$I(t,\Delta\omega) = |E(t,\Delta\omega)|^2 \quad . \tag{5}$$

The corresponding expressions, consisting of the infinite sums of the Bessel functions, have been derived for the time-dependent Mössbauer transmission [1]. In the optical region, time-dependent transmission considers the effects on phonon wings as well [27-29]. The derived expressions, as a rule, are too complicated for a direct interpretation and a computer analysis of the transmitted intensity is usually needed.

Various contributions to the time dependence of the transmitted gamma intensity can be identified in the case of thin absorber. A similar treatment has been used for the phase-modulated gamma intensity [18,30]. The inverse Fourier transform of the absorber response $R_a(\omega)$ is [31]:

$$R_a(t) = \delta(t) - b\,\exp(-i\omega_a t - \Gamma_a t/2)\,[J_1(2\sqrt{bt})/\sqrt{bt}]\,\Theta(t) \quad , \tag{6}$$

where $\delta(t)$ is the delta function and $J_1(2\sqrt{bt})$ is the first-order Bessel function of $n = 1$.

The time dependence of the transmitted amplitude (assuming $t_o = 0$) can be calculated as a convolution $E(t) = E_s(t) * R_a(t)$. The delta function in (6) allows one to express the transmitted field as a sum of the source field $E_s(t)$ and a field caused by the induced polarization of the absorber $E_a(t)$ (the absorber field):

$$E(t) = E_s(t) + E_a(t) \quad . \tag{7}$$

As follows from (7), the intensity $I(t) = |E(t)|^2$ consists of terms of three types: $|E_s(t)|^2$, $Re\{E_s(t)E_a^*(t)\}$ and $|E_a(t)|^2$. The first term is the square of the source field $|E_s(t)|^2 \sim \exp(-\Gamma_s t)$, which describes the decay of the incident intensity. For a thin ($b < \Gamma_0$) absorber $E_a(t)$ is proportional to b, i.e. to the number n of resonant nuclei in the absorber. Then $Re\{E_s(t)E_a^*(t)\} \sim -b \exp[-(\Gamma_s + \Gamma_a + b)t/2]$ describes the time dependence of the interference intensity between the incident field and the forward-scattered induced polarization. This term causes a fast decay of the transmitted intensity at the initial stage of a transient pulse (TP). The decay in intensity is fast in comparison with the one determined by the value of τ_0 and it becomes still faster when the absorber thickness is made larger. The third term, $|E_a(t)|^2 \sim b^2[\exp -(\Gamma_a + b)t]$, corresponds to the time dependence of (coherently-scattered) induced radiation. This term is proportional to b^2, i.e. to the square of the number of absorbing nuclei, n^2, and in the first approximation it describes the time-dependence of slowly-rising intensity detected in the scattering geometry.

In the case of a thick absorber the simple interpretation described above is not valid any more. Numerical calculations show that the intensity becomes oscillating after the initial fast decay, while the oscillation frequency is approximately proportional to the centre shift of the source and absorber lines, $\Delta\omega$. At the same time, an increase in $\Delta\omega$ causes a strong compression of the initial pulse. Recently, the case of a thick absorber was reconsidered in connection with the problem of a possible motion of nuclear excitation. An analysis [32] demonstrates that in the case of a significant value of the isomer shift, $\Delta\omega \geq (b\Gamma_a)^{1/2}$, and a thick absorber, $b \gg \Delta\omega$, the transmitted field consists of two pulses. The first short pulse with a very fast decay and oscillations (an analog to the Sommerfeld-Brillouin precursive pulse in optics) is followed by a slowly-rising pulse of a nuclear polariton. The latter is formed during a time interval $(\Delta\omega)^{-1}$ in the front surface of the absorber and moves through it with nearly the sound velocity (experimental values of $(0.5-0.9)10^5$ cm/s for ^{57}Fe have been obtained [33]).

In accordance with the time behaviour described the frequency dependence of the transmitted intensity for fixed short time intervals after $t_0 = 0$ (MSCG) is significantly changed when the delay time is increased. At the initial stages of a TP the MSCG line is considerably broadened in comparison with the conventional Mössbauer linewidth. When the delay time increases, the linewidth decreases gradually to the stationary value, after a while becoming still smaller. At the same time, oscillations are developing on the wings of the line, frequency regions with "negative absorption" appear and the oscillation frequency increases. The behaviour described is in a good agreement with the general ideas of the uncertainty principle about "hot" stages of the line formation in the initial decay period of the excited state and about a gradual "cooling" in the course of its decay.

Earlier experimental work has exclusively been performed on the 14.4 keV resonance of the isotope ^{57}Fe (τ_0 = 141 ns). In spite of an insufficient time resolution and a comparatively low stability of the apparatus used, the main features of the time-frequency development of transients at a fixed t_0 have been demonstrated [1,34,35]. The further improvement of the method started at the end of 70-ies and it is still continuing [9,36-40]. The time-frequency measurements have been extended to the 23.9 keV resonance (τ_0 = 26 ns) of the ^{119}Sn isotope as well [41,42].

As an example, some MSCG are presented for a ^{119}Sn single-line source-absorber combination (both in the form of $BaSnO_3$) [42], which clearly demonstrate the above-described phenomena (Figure 1).

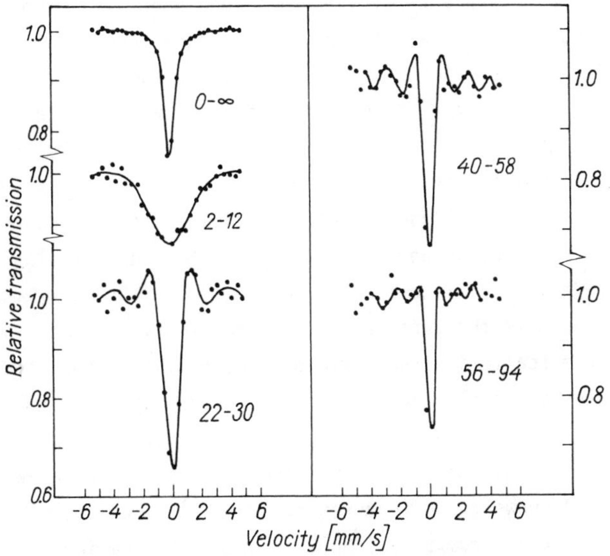

Figure 1: MSCG in ^{119}Sn transmission [42] at various delay time intervals t_1-t_2 (ns) with respect to $t_0 = 0$, i.e. the moment of the detection of a 26 keV X-ray quantum. The gamma source is $Ba^{119m}SnO_3$; the absorber, $Ba^{119}SnO_3$ ($b \approx 0.5\ \Gamma_0$). The recording conditions for the stationary spectrum 0 - ∞ and MSCG were identical

2.2. Scattering

Resonant absorber nuclei excited by the radiation relax by emitting gamma-quanta or conversion electrons with accompanying X-ray quanta and Auger electrons. The detection of these secondary radiations opens a possibility for transient studies in scattering geometry. The theory of the time-dependent Mössbauer scattering has been developed in [35,43], while the first experiments on ^{57}Fe have demonstrated a qualitative agreement between the theory and experiments.

The formation of scattering transients can be described by using a simplified model, neglecting thereby angular dependences and the relative contribution by the secondary radiation. Assume incident source field (2) to be perpendicular to a thin scatterer foil of the thickness d, where $b = \beta d$. The response of a single resonant nucleus to the exciting field is proportional to $(\omega - \omega_a - i\Gamma_a/2)^{-1}$ (see (3)). The effect of the scatterer thickness on the source field at the depth x is taken into account by a complex exponential factor $\exp[i\beta x(\omega - \omega_a - i\Gamma_a/2)]$. Then the scattering amplitude of a resonant nucleus is described in the form [43]:

$$\phi(\omega,x) \sim (\omega - \omega_a - i\Gamma_a/2)^{-1} \exp[i\beta x(\omega - \omega_a - i\Gamma_a/2)]\, E_s(\omega) \quad . \qquad (8)$$

The time-frequency dependence of the scattered intensity is obtained by integrating (8) over ω, squaring the result and integrating it over x. The general expressions found in [35,43] are rather cumbersome. At the same time, for the case of precise resonance ($\Delta\omega = 0$), $I(t) \sim t^2 \exp(-\Gamma_a t)$ for an extremely thin scatterer ($b \to 0$) and $I(t) \sim t \exp(-\Gamma_a t)$ for an infinitely thick scatterer, while I(t) for the real thickness can be found between these extremes. In the case of $\Delta\omega = 0$ the intensity of the scattering transients rises relatively slowly after $t = 0$, reaches a maximum value in the time interval between τ_0 and $2\tau_0$ and then decays slowly. An increase in the line shift $\Delta\omega$ moves the intensity maxima towards $t = 0$, makes the decaying curve oscillate and compresses the duration of the transient. In comparison with transmission transients the scattering ones develop considerably slower and are characterized by relatively low oscillation amplitudes. Accordingly, contrary to transmission, the scattering MSCG lines usually show no oscillations at the wings. The most characteristic features are the changes in the linewidth and in the peak intensity with the delay. Figure 2 shows these features for the ^{119}Sn MSCG measured by detecting the conversion electrons from a $Ba^{119}SnO_3$ scatterer for various delay times. The linewidth for the maximum delay (120-144 ns) is smaller than $2\tau_0$.

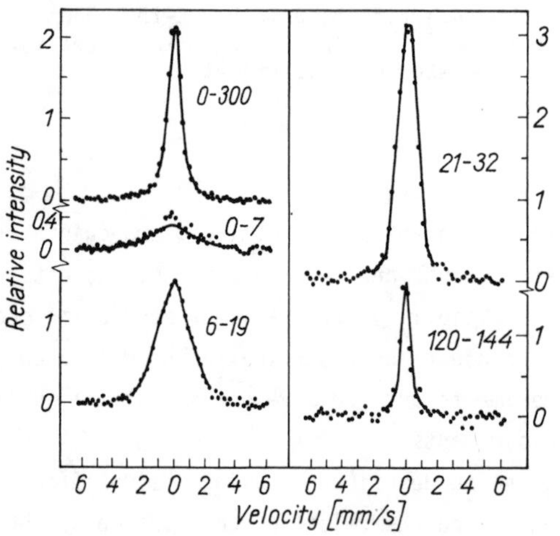

Figure 2: The MSCG of ^{119}Sn [42] measured in the scattering geometry for various delay time intervals t_1-t_2 (ns). The MSCG for 0-300 ns practically coincides with the stationary Mössbauer spectrum. The gamma source was $Ba^{119}SnO_3$; the scatterer $Ba^{119}SnO_3$ ($b \approx 0.125\ \Gamma_0$) was incorporated in a scintillation resonance detector for recording conversion electrons

The results of not numerous experiments on scattering transients [35,42-45] show a good quantitative agreement of theoretical and experimental time dependences for the case of precise resonance only, while for $\Delta\omega \neq 0$ considerable discrepancies appear even in the case of a thin scatterer.

In spite of the efforts to explain these discrepancies by taking into account the nonresonant efficiency of the detector [46] or inhomogeneous broadening [44] etc., their origin is not fully understood yet. Therefore, further theoretical and experimental work is needed. At the same time, the experiments on scattering transients by using the Mössbauer coincidence are extremely troublesome due to the intrinsic limit of the method on the activity of the gamma source. These experiments can be greatly facilitated by another type of transient studies - by the method of stepwise phase modulation of the resonant gamma radiation.

3. Phase-Switching Transients of Gamma Radiation

3.1. Transmission

TPh in the case of the phase switching of gamma radiation can be explained by the following model [18,30,47]. Consider the above-described source-absorber combination with single unsplit lines of field amplitude (1) and response (3), respectively. At the time moment T the gamma source as a whole is mechanically instantaneously shifted to a distance Δx, which results in a stepwise shift of the radiation phase $a\Theta(t-T)$, where $a = \Delta x/\lambda$ and $\Lambda = 2\pi\lambda$ is the wavelength of the resonant gamma radiation (for ^{57}Fe $\lambda = 0.137$ Å). We assume the conditions of resonance to be stationary, i.e. $\Delta\omega = \mathrm{const}$ before and after the phase shift. In this case the intensity of the radiation passing through the absorber, $N(t<T)$, is also stationary. The time dependence is recordered with respect to the time moment T, while the random time moments t_0 are not registered.

The amplitude of the phase-modulated electric field of the resonant gamma radiation is obtained by adding to (1) a phase factor [18]:

$$E_s(t) \sim E_s(t - t_0)\exp[ia\Theta(t - T)] \quad . \tag{9}$$

Field (9) can be presented in the form of the sum of wave packets [47] emitted at random time moments t_0,

$$E_r(t - t_0) = E_s(t - t_0)\exp[ia\Theta(t_0 - T)]$$

and at the moment of the phase step, T,

$$E_i(t - T) = (e^{ia} - 1)\,E_s(t - T)\,E_s(T - t_0) \quad .$$

Consequently, $E_s(t) = E_r(t - t_0) + E_i(t - T)$, whereby in the absence of an absorber no step occurs in the field intensity at the time moment $t = T$ as the result of the interference of these two terms. Various modifications of the amplitude and phase of $E_r(t - t_0)$ and $E_i(t - T)$, caused by absorber response (3), induce TP in the intensity of the radiation passing through the absorber [47]:

$$I(t,\Delta\omega) = [N(t) - N(t<0)]/N$$

$$= 2f_s\, e^{-\Gamma_a t}\, [(1 - \cos a)\, \mathrm{Re}\{A\circ B\} - \sin a\, \mathrm{Im}\{A\circ B\}]\, \Theta(t) \quad ,$$

where

$$A = \sum_{n=0}^{\infty} J_n(2\sqrt{bt})\, (-1)^n\, (t/b)^{n/2}\, [i\Delta\omega + (\Gamma_s - \Gamma_a)/2]^n \quad ,$$

$$B = \sum_{n=1}^{\infty} J_n(2\sqrt{bt})\, (-1)^{n+1}\, (b/t)^{n/2} [i\Delta\omega + (\Gamma_s + \Gamma_a)/2]^{-n} \quad . \qquad (10)$$

Here $T = 0$, N_∞ is the stationary intensity at $\Delta\omega \to \infty$. A detailed derivation of formula (10) has been performed for the frequency [47] and time [30] domains. The obtained version of formula (10) is too complicated to allow any detailed interpretation. However, like in the case of switch-in TP, for a thin absorber the time dependence of intensity can also be presented in the form of the sum of three terms [30]. Here, too, one can distinguish the interference term $|E_s(t)\, E_a^*(t)| \sim b$ which quickly reacts to the phase jump, and the squared term $|E_a(t)|^2 \sim b^2$ caused by scattering. The term $|E_s(t)|^2$ represents a constant background (instead of $\exp(-\Gamma_s t)$ for switch-in TP). Such situation is understandable as here occurs averaging over the times t_0. In general, time dependences and MSCG are different for both cases. This can be clearly seen from a comparison of time dependences at $\Delta\omega = 0$ and $\Gamma_a = \Gamma_s = \Gamma_0$. In the case of a fixed T the TP of the phase-switching is of the form [47]:

$$I_T(t) \sim \Theta(t)\, e^{-\Gamma_0 t}\, J_0(2\sqrt{bt}) \sum_{n=1}^{\infty} (-1)^{n+1}\, (b/t)^{n/2}\, J_n(2\sqrt{bt}) \quad , \qquad (11)$$

while in the case of the time reference t_0 [1] the switch-in TP is described by the expression:

$$I_{t_0}(t) \sim \Theta(t)\, e^{-\Gamma_0 t}\, [J_0(2\sqrt{bt})]^2 \quad . \qquad (12)$$

It follows that if the thickness b is small, the decay of the intensity after the initial jump is somewhat slower than $I_{t_0}(t)$ and that the intensities practically coincide when the thickness of the absorber grows.

An analysis of formula (10) shows that the stepwise shift of the phase is accompanied by a considerable jump in the intensity of the resonant gamma radiation, passing through the absorber. In the case of an exact resonance, $\Delta\omega = 0$, a thick absorber and $a = (2n + 1)\pi$, the peak intensity of TP is four times higher than the stationary effect value, ε_0, for the source-absorber combination used. Still higher jumps, reaching up to $8\varepsilon_0$, should be observed in an off-resonance case. The larger the thickness b is the more rapid is the nearly-exponential decay following the jump. The decay ends with oscillations whose amplitudes grow and shift towards $t = 0$ with the growth of b. At $a = 2n\pi$ TP should not be observed, as the intensity of the jump is proportional to $(1 - \cos a) \to 0$.

The intensity and form of TP are strongly dependent on the value and sign of the phase step a and on the shift of the centres of the lines $\Delta\omega$. The TP intensity is rather sensitive to the centre shift (especially in the case of $a < \pi$), which enables a precise measurement of extremely small shifts of the Mössbauer lines within $|\Delta\omega| \leq$ $\leq (1 \ldots 2)\Gamma_0$ [48,49].

The feasibility of TP on the phase switching of the Mössbauer gamma radiation has been experimentally proved on the 93.3 keV gamma resonance of ^{67}Zn isotope ($\tau_0 = 13.2$ μs) [18,49] and on a more applicable 14.4 keV Mössbauer resonance of ^{57}Fe isotope ($\tau_0 = 141$ ns) [47,48,50,51]. In both cases, the rapid mechanical shifts of the source (or absorber), inducing periodical phase jumps of the gamma source, were implemented by means of quartz piezotransducers. Let us consider the results obtained for ^{57}Fe, which have been studied in more detail.

Owing to the short lifetime of ^{57}Fe, the requirements imposed on the modulator parameters for fast response and the lack of distortions in the motion are rather strict. Until now these methodical problems of the experiment have not been fully solved and, therefore, to analyse the experimental results on TP modifications should be made in the theory of phase modulation including the multitude of disturbing factors. The relevant problems have been considered and formulae obtained, which describe the action of the following real factors of the experiment: (i) the mechanical shift (phase shift) of the source takes place in a finite time by some law a(t) [30,47], (ii) a spread of the shifts exists in the volume of the gamma source [47,52,53].

To record the time dependences and MSCG with respect to the piezotransducer-induced stepwise shifts a Mössbauer spectrochronograph was used, which is a spectrometer with a time analyser based on a time-amplitude converter [37,54]. In the case of a stepwise modulation, the latter enables the measurements of three different dependences of gamma radiation intensities (in the case of transmission geometry) or conversion electrons (in the case of scattering geometry): (i) time dependences for various excitation amplitudes $U \sim a$ and a constant centre shift $S \sim \Delta\omega = \text{const}$, (ii) the same for various $S \sim \Delta\omega$ and a constant amplitude $U = \text{const}$ and (iii) MSCG - frequency dependences of the intensity for various time intervals $(t, t + \Delta t)$ with respect to the front of exciting pulses and $U = \text{const}$. As a rule, only the TP corresponding to the jump of the gamma source towards the absorber/scatterer were recorded.

A block-diagram of the experimental setup in the scattering geometry [54] is depicted in Figure 3. The time resolution of the apparatus is 4.8 ns for transmission and 28 ns for scattering. As the actual duration of modulational jumps is essentially longer (∿80 ns), the effect of the apparatus resolution can be neglected in the analysis.

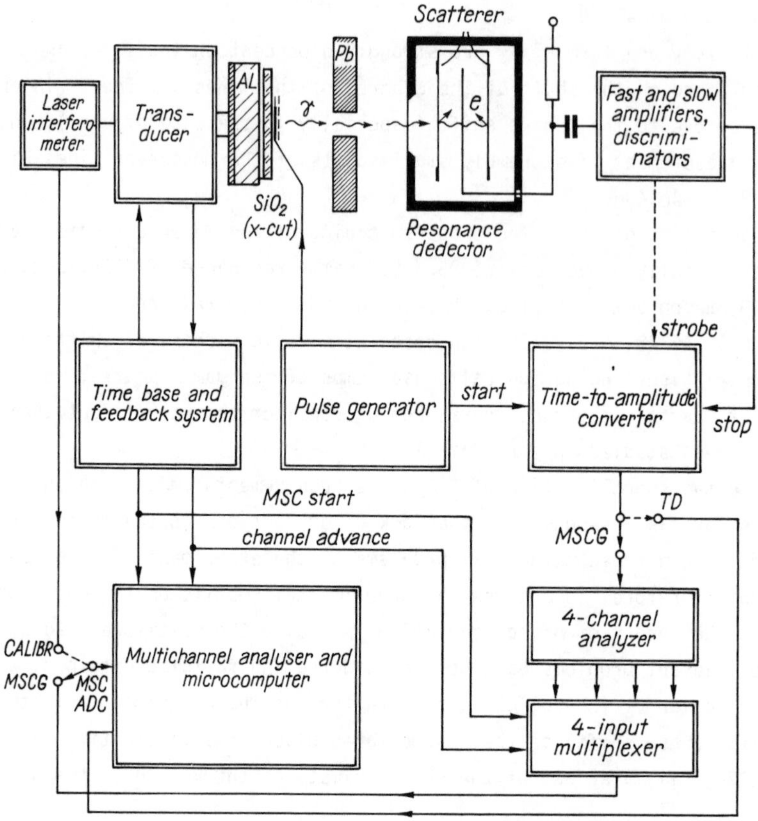

Figure 3: A block-diagram of the Mössbauer spectrochronograph for measuring the time dependences (TD) and Mössbauer spectrochronograms (MSCG) of the transient pulses (TP) of scattering on the phase switching of the gamma radiation of ^{57}Fe [54]. For measuring the TP of transmission the resonance detector is replaced by an absorber, $K_4{}^{57}Fe(CN)_6 \cdot 3H_2O$, and a scintillation detector NaI(Tl) with a fast photomultiplier

In Figure 4, the transmission TP of ^{57}Fe are presented for a number of values of the exciting pulse amplitudes, U, in the case of $S = 0.00 \pm 0.01$ mm/s [47]. The phase jump induces a synchronous jump in the intensity. This synchronous jump grows with the growth of amplitude up to $U \simeq 25$ V, remaining then practically constant $I_+ = 2.8$ (with

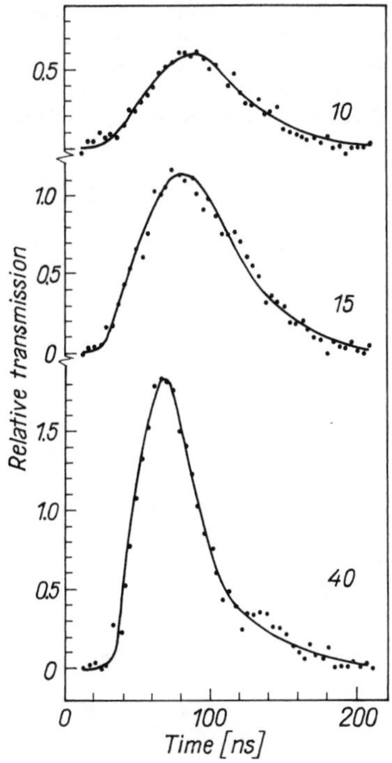

Figure 4: TP on the phase switching of resonant gamma radiation of ^{57}Fe in transmission for various amplitudes of exciting pulses, U (V) and $S \sim \Delta\omega = 0$. The intensity I_+ is expressed in the units of the stationary effect value ε_0. Solid curves are the theoretical dependences considering the finite duration and distribution of phase jump values [47]. The source $^{57}Co(Pd)$ is cemented on a x-cut quartz piezotransducer

respect to $I(t<0)$). On the scale of the stationary effect $I_+ = 1.84\ \varepsilon_0$. With the rise in intensity the TP is shortened and their maxima are shifted towards $t = 0$. The risetime of TP decreases from 40 ± 2 ns at 10 V to 27.5 ± 2.0 ns at 40 V and FWHM of TP, from 80 ± 4 ns (U = 10 V) to 48 ± 4 ns (U = 40 V). The rise of the TP intensity above ε_0 and the decrease of their duration to less than τ_0 confirms the realization of TPh on ^{57}Fe isotope. Recent improvements in the transducer design show that much shorter risetimes (22 ns) and pulsewidths (36 ns) can be achieved [51].

One can infer from the curves that the shape and characteristics of TP are not quantitatively explainable within the framework of pure stepwise modulation. At the same time, a good agreement of the theory (points in Figure 4) and experiment (solid curves in Figure 4) is obtained by assuming a 80 ns-duration linear mechanical source shift and a cosine radial distribution of these shifts over the foil surface of the source with the amplitude Δx_m in the centre and 0.5 Δx_m on the edges [47] ($\Delta x_m \sim U$).

The shape, intensity and sign of TP are strongly dependent on the changes of the value and sign of the centre shift $S \sim \Delta\omega$ at U = const (Figure 5). In accordance with the theory the maximum intensity is observed not at S = 0, where $I_+(0) = 1.07\ \varepsilon_0$, but at about S = 0.10 mm/s, where $I_+ = 1.38\ \varepsilon_0$. The increase of $|S|$ leads to a nearly linear decrease of I_+, to the rise of oscillations, to the change of the sign of I_+ and to a shift of the TP edge towards t = 0.

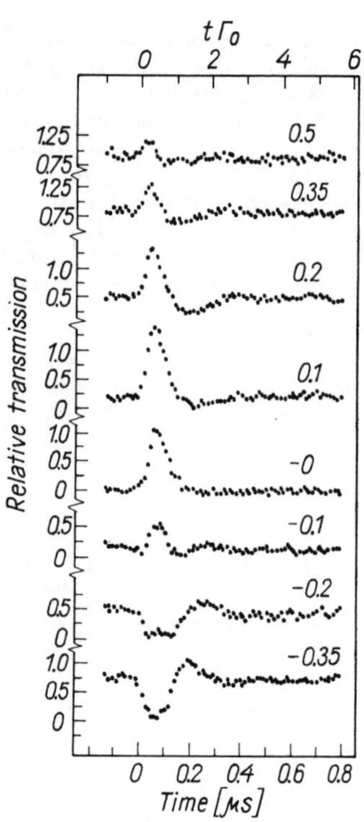

<u>Figure 5</u>: TP of ^{57}Fe on the phase switching in transmission for various centre shifts S (mm/s) and U = 15 V. The intensity I_+ is given in the units of ε_o. The dependence $I_+(S, t < 0)$ describes the stationary Mössbauer spectrum.

A comparison of the dependences $I_+(U)$ (Figure 6a) and $I_+(S)$ (Figure 6b) also points to a good agreement between the experiment and the modified theory

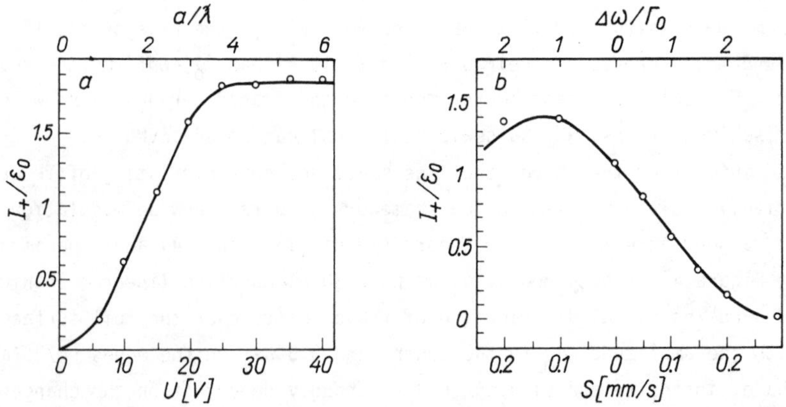

<u>Figure 6</u>: The dependence of the peak intensities of the TP of ^{57}Fe in transmission on the amplitude of exciting pulses, $U \sim a_m$ for S = 0 (left) and on the centre shift S for U = 15 V (right). The solid lines correspond to theoretical calculation [47]

The complexity of the time-frequency dependence is also confirmed by the shape of the MSCG (Figure 7,II) measured at various time intervals Δt with respect to $t = 0$ (Figure 7,I). At various stages of the TP development the regions of the so-called "negative absorption", line dispersion and the shift of its minimum can be observed.

The results of the investigation of TP in transmission are important from the methodical point of view. The possibility of forming periodic high-energy resonant 14.4 keV gamma radiation pulses of ^{57}Fe of the duration less than τ_0 extends the application of time-domain methods in Mössbauer spectroscopy. This applies especially to the possibilities of the investigation of the effects of various external pulse excitations of samples. Besides, TP can successfully be utilized in rather precise measurements of the amplitudes and velocities of mechanical shifts within 0-1 Å and ±150 μm/s, respectively, and also for detecting extremely small shifts of resonant lines.

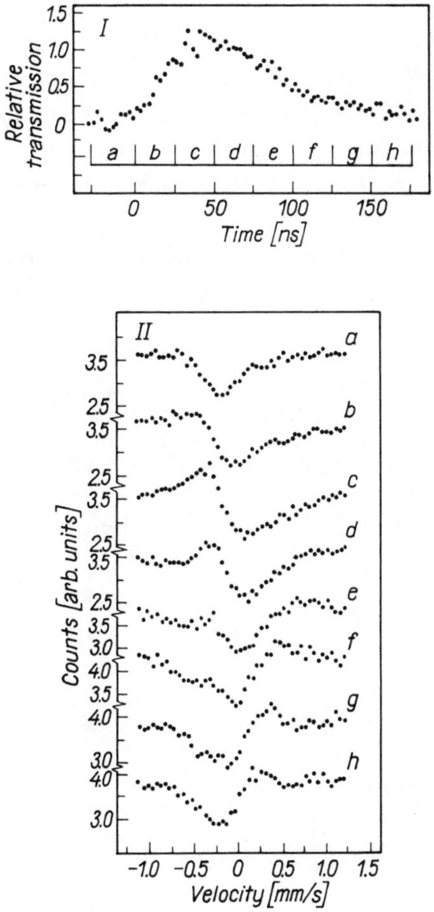

Figure 7: TP in transmission for S = 0, U = 15 V (I); the 26 ns time intervals (a ÷ h) are shown for which MSCG were measured (II)

3.2. Scattering

The theory of TP on the phase switching of the Mössbauer gamma radiation has been developed in [55]. Like in the case of the TP of radiation switch-in, here the scattering amplitude of the nucleus is considered in the form of (8) at the depth x. It is assumed that on stepwise phase modulation the field of the source (9) is perpendicularly incident on the scatterer's foil surface of the thickness d ($b = \beta d$). The time dependence of the scattered intensity was calculated by integrating the square of the Fourier transform of (8) $\phi(t,x,t_0)$ over x and t_0. In the thin scatterer approximation ($b \ll \Gamma_a$), for the intensity of conversion electrons, X-ray quanta, etc. the following expression is obtained [55]:

$$I(t,\Delta\omega) \sim 2b\Gamma_a \,\mathrm{Re}\{\frac{1}{\gamma_+(\gamma_+ + \gamma_-)} - (1 - e^{ia}) \frac{e^{-\gamma_+ t}(1 - e^{-\gamma_- t})}{\gamma_+\gamma_-} \Theta(t)\} \quad , \qquad (13)$$

where $\gamma_\pm = (\Gamma_a \pm \Gamma_s + b)/2 + i\Delta\omega$. From (13) it follows that in the case of a very thin absorber ($b \to 0$), exact resonance and equal linewidths ($\Gamma_s = \Gamma_a$) the time dependence approaches $I(t) = -t\, \exp(-\Gamma_a t)$. Remember for comparison that in the case of the switch-in TP the corresponding dependence was $t^2 \exp(-\Gamma_a t)$. In scattering, TP are to a considerably less extent influenced by the distortions of phase modulation than in transmission. Below we shall regard their main features on the example of experiments on ^{57}Fe [54,55].

In Figure 8, a series of time dependences of the emission of conversion electrons from the thin ^{57}FeAl layer under the action of a stepwise-modulated Mössbauer gamma

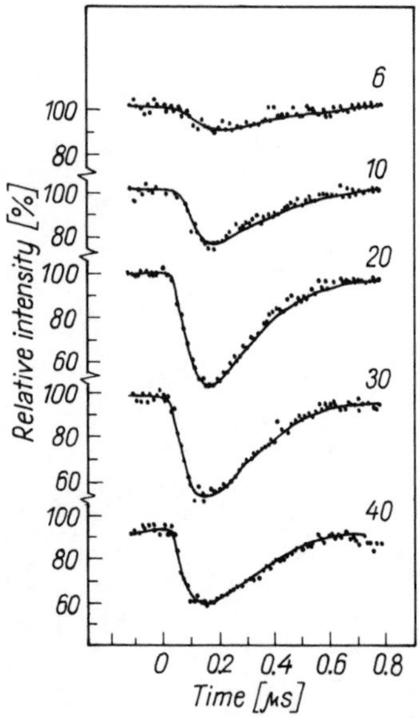

Figure 8: The phase-switching TP of ^{57}Fe gamma radiation in the scattering geometry for various amplitudes $U \sim a$ (V) and $S = 0$. The scatterer ^{57}FeAl is included in a resonant detector of 7.3 keV conversion electrons. Solid curves - theoretical dependences (13) for $\Gamma_s = \Gamma_a = \Gamma_0$, $\Delta\omega = 0$ and $b = 0.25\ \Gamma_0$

radiation is presented. TP were measured at various amplitudes U in exact resonance S = 0. With the growth of U the intensity increases up to 50% (U = 25 V) with respect to the one before the phase jump and then decreases up to 40% at 40 V. The intensity reaches its minimum about 125 ns after t = 0 and FWHM of TP makes ∿270 ns. In scattering, TP are considerably more inert than in transmission. The measured dependences are in good agreement with those calculated by (12) at $\Gamma_s = \Gamma_a = \Gamma_o$, $\Delta\omega = 0$ and $b = 0.25\ \Gamma_o$ (solid curves). In case $U \geq 30$ V, the best agreement is obtained for somewhat smaller intensity values than those following from formula (12). Here, evidently, modulation-induced low-frequency distortions of the transducer motion become manifest.

The dependence of the shape of the scattering TP on S at U = 15 V is shown in Figure 9. The dependence of the stationary intensity level before the phase jump I(t < 0) on S is described as an ordinary Mössbauer spectrum with $\Gamma_a \simeq 2\Gamma_o$ and $\Gamma_s = 1.1\ \Gamma_o$. Further investigations are needed to clarify why the strong line broadening, peculiar to the scatterer material ($\gamma_a \simeq \Gamma_o$), is not manifest in time dependence. TP are positive in the region of shifts $S \geq 0.12$ mm/s and negative in the rest of the region, whereby their intensities and duration times decrease with the increase of $|S|$. In the region S < 0, oscillatory behaviour is observed: a negative pulse is followed by a zero-crossing and a positive overshoot. When moving off the resonance, a strong shift of the front edge and the maximum/minimum point of the TP towards $t \to 0$ occurs. The front edge of the TP shifts more than 150 ns, while the centre shift increases up to 0.6 mm/s = 6 Γ_o. For the scattering TP the thin scatterer foil behaves like a delay line controlled by tuning over $S \sim \Delta\omega$ (note that in vacuum gamma radiation traverses 45 m during 150 ns).

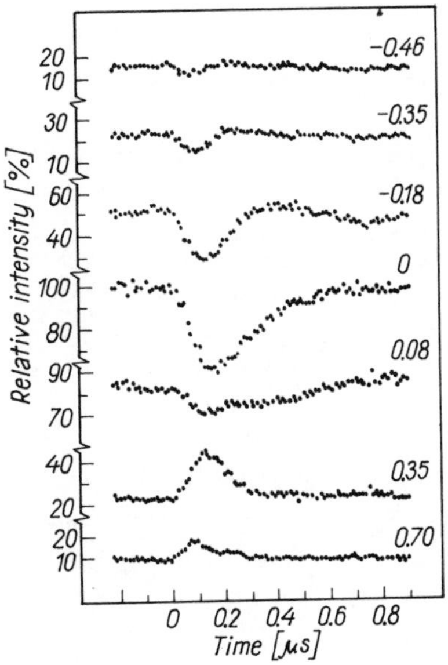

Figure 9: TP of ^{57}Fe scattering for various centre shifts S (mm/s) and U = 15 V. Intensity was normalized to 100% at S = 0 and t < 0. The dependence of the intensity up to the phase jump I(t < 0) is described by a stationary Mössbauer spectrum in the scattering geometry with $\Gamma_s = 1.1\ \Gamma_o$, $\Gamma_a = 2\Gamma_o$ and $b = 0.25\ \Gamma_o$

In Figure 10,I, the time intervals at various development stages of TP are depicted, for which MSCG were measured (Figure 10,II). Before the modulational jump the MSCG coincides with the stationary spectrum. At $t > 0$ one can observe the time-dependent asymmetry and the line broadening as well as a shift of the maximum and its centre towards positive Doppler velocities. In general, the measured MSCG are in good agreement with the time dependences in Figure 9.

In conclusion note that the investigations of TPh in resonant gamma radiation essentially extend the possibilities of Mössbauer spectroscopy. The elucidation of

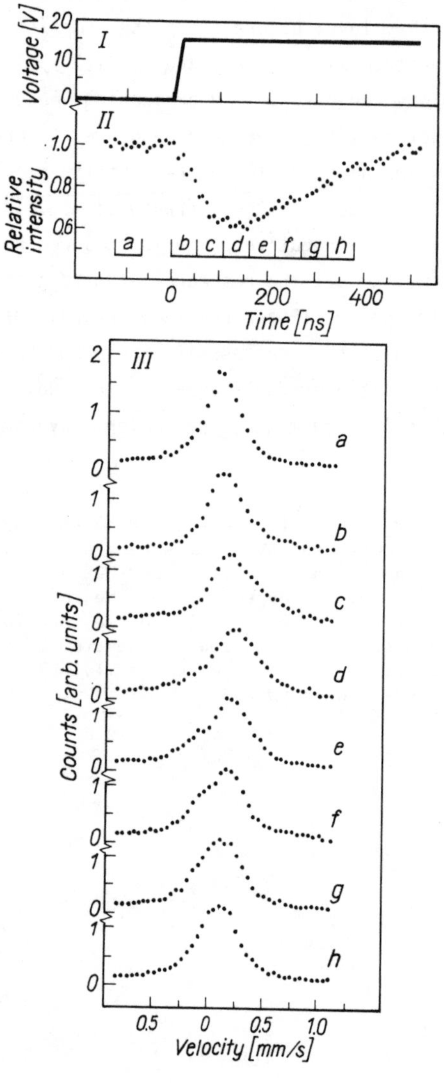

Figure 10: I - the voltage pulse U(t) exciting the x-cut quartz piezotransducer; II - the scattering TP with selected 45 ns time intervals (a - h); III - the corresponding MSCG in scattering

the fundamental characteristics of TPh may be helpful in the clarification of various relaxation processes in solids, mechanisms of Mössbauer line broadening, etc. Metrological applications are also possible.

An analysis of theoretical studies and the experimental results indicate that the method of the phase switching of resonant gamma radiation is especially favourable in the investigations of the time-frequency characteristics of scattering, diffraction, etc. The versatility of a similar method has been successfully demonstrated on measuring the time-dependences of the Mössbauer diffraction [56,57]. Note that the inclusion of the TPh of the Mössbauer gamma radiation is also important in the interpretation of recent promising experiments on exciting the Mössbauer levels by ultrashort pulses of synchrotron radiation [58,59].

The author is indebted to K. Rebane for his support and continuous interest in this work, to J. Jõgi, R. Koch and H. Raudsepp for their help in experiments. Thanks are also due to M. Haas and V. Hizhnyakov for useful discussions.

References

1 F.J. Lynch, R.E. Holland, M. Hamermesh: Phys. Rev. 120, 513 (1960)
2 G.J. Perlow: Phys. Rev. Lett. 40, 896 (1978)
3 J.E. Monahan, G.J. Perlow: Phys. Rev. A20, 1499 (1979)
4 K.K. Rebane, V.V. Hizhnyakov: Opt. Spektrosk. 14, 362 (1963); 14, 491 (1963) [English transl.: Opt. Spectrosc. USSR 14, 193 (1963); 14, 262 (1963)]
5 K.K. Rebane: Opt. Spektrosk. 16, 594 (1964) [English transl.: Opt. Spectrosc. USSR 16, 324 (1964)]
6 K.K. Rebane: Elementarnaya teoriya kolebatel'noj struktury spektrov primesnykh tsentrov kristallov (Nauka, Moskva 1968) [English transl.: Impurity Spectra of Solids (Plenum Press, New York 1970)]
7 V.P. Alekseev, V.I. Gol'danskij, V.E. Prusakov, A.V. Nefed'ev, R.S. Stukan: Pis'ma Zh. Eksp. Teor. Fiz. 16, 65 (1972) [English transl.: JETP Lett. 16, 43 (1972)]
8 G.R. Hoy, P.P. Wintersteiner: Phys. Rev. Lett. 28, 877 (1972)
9 R. Grimm, P. Gütlich, E. Kankeleit, R. Link: J. Chem. Phys. 67, 5491 (1977)
10 R. Koch, E. Realo: Pis'ma Zh. Eksp. Teor. Fiz. 30, 716 (1979) [English transl.: JETP Lett. 30, 679 (1979)]
11 T. Kobayashi, T. Kitahara: J. Phys. Colloq. C41, 239 (1980)
12 H. Berlin, J. Schmand: J. Phys. Colloq. C41, 135 (1980)
13 E. Kankeleit: Z. Phys. A275, 119 (1975)
14 E. Kankeleit, A. Körding: J. Phys. Colloq. C37, 65 (1976)
15 A.M. Afanas'ev, V.D. Gorobchenko: Phys. Status Solidi b76, 465 (1981)
16 A. Gupta, K.R. Reddy: Phys. Lett. A60, 58 (1977)
17 P. Helistö, T. Katila, W. Potzel, K. Riski: Phys. Lett. A85, 177 (1981)
18 P. Helistö, E. Ikonen, T. Katila, K. Riski: Phys. Rev. Lett. 49, 1209 (1982)
19 A.Z. Genack, D.A. Weitz, R.M. Macfarlane, R.M. Shelby, A. Schenzle: Phys. Rev. Lett. 45, 438 (1980)
20 A.Z. Genack, R.O. Brickman, A. Schenzle: Appl. Phys. B28, 276 (1982)
21 F. Shimizu, K. Umezu, H. Takuma: Phys. Rev. Lett. 47, 825 (1981)
22 F. Shimizu, K. Shimizu, H. Takuma: Phys. Rev. A28, 2248 (1983)
23 A. Freiberg, P. Saari: IEEE J. QE-19, 622 (1983)
24 E.L. Hahn: Phys. Rev. 77, 297 (1950)
25 E.L. Hahn: Phys. Rev. 80, 580 (1950)
26 N. Bloembergen, E.M. Purcell, R.V. Pound: Phys. Rev. 73, 679 (1948)

27 V.V. Hizhnyakov: Tech. Rep. of JSSP, Univ. of Tokyo, Ser. A, No. 860 (1977)
28 I.K. Rebane, A.L. Tuul, V.V. Hizhnyakov: Zh. Eksp. Teor. Fiz. 77, 1302 (1979)
29 I.K. Rebane, V.V. Hizhnyakov: Eesti NSV Tead. Akad. Toim. Füüs. Matem. (Proc. Estonian SSR Acad. Sci.) 30, 1 (1981) (in Russian)
30 E. Ikonen, P. Helistö, T. Katila, K. Riski: Phys. Rev. A32, 2298 (1985)
31 Yu. Kagan, A.M. Afanas'ev, V.G. Kohn: J. Phys. C12, 615 (1979)
32 E. Realo, M. Haas, V. Hizhnyakov: Eesti NSV Tead. Akad. Toim. Füüs. Matem. (Proc. Estonian SSR Acad. Sci.) 36, 113 (1987) (in Russian)
33 M. Haas, V. Hizhnyakov, E. Realo: Phys. Lett. A124, 370 (1987)
34 C.S. Wu, Y.K. Lee, N. Benzer-Koller, P. Simms: Phys. Rev. Lett. 5, 432 (1960)
35 W. Neuwirth: Z. Phys. 197, 473 (1966)
36 T. Kobayashi, K. Fukumura: Nucl. Instrum. and Meth. 166, 257 (1979)
37 R. Koch, E. Realo: Eesti NSV Tead. Akad. Toim. Füüs. Matem. (Proc. Estonian SSR Acad. Sci.) 28, 374 (1979) (in English)
38 Zs. Kajcsos, M. Alflen, H. Spiering, P. Gütlich: in Abstracts of the International Conference on the Applications of Mössbauer Effect (Leuven 1985), No. 12.21
39 K.P. Purcell, M.P. Edwards, B.P. Curnutte, J.S. Eck: Rev. Sci. Instrum. 56, 108 (1985)
40 R. Albrecht, M. Alflen, P. Gütlich, Zs. Kajcsos, R. Schulze, H. Spiering, F. Tuszek: Nucl. Instrum. Meth. A257, 209 (1987)
41 J. Carmeliet, S. Lejeune: Nucl. Instrum. Meth. 86, 93 (1970)
42 R. Koch, E. Realo: Eesti NSV Tead. Akad. Toim. Füüs. Matem. (Proc. Estonian SSR Acad. Sci.) 30, 171 (1981) (in English)
43 P. Thieberger, J.A. Moragues, A.W. Sunyar: Phys. Rev. 171, 425 (1968)
44 H. Drost, H.V. Lojewski, K. Palow, R. Wallenstein, G. Weyer: in Proceedings of the Vth International Conference on Mössbauer Spectroscopy, Part 3, ed. by M. Hucl, T. Zemčik (Praha 1975) p. 713
45 E.I. Vapirev, P.S. Kamenov, V. Dimitrov, D.L. Balabanski: Nucl. Instrum. Meth. 219, 376 (1984)
46 D.L. Balabanski, P.S. Kamenov, V. Dimitrov, E. Vapirev: Nucl. Instrum. Meth. B16, 100 (1986)
47 E.H. Realo, M.A. Haas, J.Ä. Jõgi: Zh. Eksp. Teor. Fiz. 88, 1864 (1985) [English transl.: Sov. Phys. - JETP 61, 1105 (1985)]
48 R. Koch, E. Realo, K. Rebane, J. Jõgi: in Abstracts of the International Conference on the Applications of Mössbauer Effect (Alma-Ata 1983) p. 389
49 P. Helistö, E. Ikonen, T. Katila, W. Potzel, K. Riski: Phys. Rev. B30, 2345 (1984)
50 E.H. Realo, K.K. Rebane, M.A. Haas, J.Ä. Jõgi: Pis'ma Zh. Eksp. Teor. Fiz. 40, 477 (1984) [English transl.: JETP Lett. 40, 1309 (1984)]
51 P. Helistö, E. Ikonen, T. Katila: Phys. Rev. B34, 3458 (1986)
52 R.R. Koch, E.H. Realo: Prikladnaya yadernaya spektroskopiya, Part 13 (Atomizdat, Leningrad 1984) p. 93
53 E.H. Du Marchie van Voorthuysen, G.L. Zhang, H. De Waard: Phys. Rev. A30, 2356 (1984)
54 E.H. Realo, J.Ä. Jõgi, S.I. Reiman: Izv. AN SSSR Ser. Fiz. 50, 2334 (1986)
55 P. Helistö, E. Ikonen, T. Katila, J. Jõgi, E. Realo, S. Reiman: Helsinki Univ. of Technology, Report TKK-F-A571 (Otaniemi 1985); Eesti NSV Tead. Akad. Toim. Füüs. Matem. (Proc. Estonian SSR Acad. Sci.) 35, 308 (1986) (in English)
56 G.V. Smirnov, Yu.V. Shvyd'ko, E. Realo: in Abstracts of the International Conference on the Applications of Mössbauer Effect (Alma-Ata 1983) p. 407
57 G.V. Smirnov, Yu.V. Shvyd'ko, E. Realo: Pis'ma Zh. Eksp. Teor. Fiz. 39, 33 (1984) [English transl.: JETP Lett. 39, 41 (1984)]
58 E. Gerdau, R. Rüffer, H. Winkler, W. Tolksdorf, C.P. Klages, J.P. Hannon: Phys. Rev. Lett. 54, 835 (1985)
59 U. van Bürck, R.L. Mössbauer, E. Gerdau, R. Rüffer, R. Hollatz, G.V. Smirnov, J.P. Hannon: Phys. Rev. Lett. 59, 355 (1987)

Index of Contributors